DISEASES OF FOREST TREES AND THEIR MANAGEMENT

The science of Forest Pathology is a complex amalgamation of forestry, plant pathology, ecology, and biology. The book is intended to provide comprehensive introduction to the important aspects of the field of forest pathology and tree diseases. The book is arranged in two major parts. The fundamental chapters, present forest diseases, pathogens, epidemics, and management that is applicable to all forest trees. The applied chapters on the individual crops that are grouped alphabetically present information on the symptoms, pathogen and integrated management of major diseases of forest trees. It was designed to give a broad overview of the field of forest pathology but with sufficient detail that they will be able to assess their specific role as practicing forestry professionals.

Dr. S. Parthasarathy is Assistant Professor (Plant Pathology), Amrita School of Agricultural Sciences, Amrita Vishwa Vidyapeetham, Coimbatore. He has completed his Ph.D. from Tamil Nadu Agricultural University, Coimbatore.

Dr. G. Thiribhuvanamala is Associate Professor in Department of Plant Pathology, Tamil Nadu Agricultural University, Coimbatore. She has completed her Ph.D. from Tamil Nadu Agricultural University, Coimbatore.

Dr. P. Muthulakshmi is Associate Professor (Plant Pathology) in the Department of Plant Pathology, Tamil Nadu Agricultural University, Coimbatore.

Dr. K. Angappan is a Professor (Plant Pathology) in the Department of Plant Pathology, Tamil Nadu Agricultural University, Coimbatore.

DISEASES OF FOREST TREES AND THEIR MANAGEMENT

S. Parthasarathy
Amrita School of Agricultural Sciences
Amrita Vishwa Vidyapeetham, Coimbatore

G. Thiribhuvanamala
Department of Plant Pathology
Tamil Nadu Agricultural University, Coimbatore

P. Muthulakshmi
Department of Plant Pathology
Tamil Nadu Agricultural University, Coimbatore

K. Angappan
Department of Plant Pathology
Tamil Nadu Agricultural University, Coimbatore

NPH

NEW DELHI - 110085 (INDIA)

First published 2021
by CRC Press
2 Park Square, Milton Park, Abingdon, Oxon, OX14 4RN

and by CRC Press
6000 Broken Sound Parkway NW, Suite 300, Boca Raton, FL 33487-2742

CRC Press is an imprint of Informa UK Limited

British Library Cataloguing-in-Publication Data
A catalogue record for this book is available from the British Library

Library of Congress Cataloging-in-Publication Data
A catalog record has been requested

ISBN: 978-1-032-00370-2 (hbk)
ISBN: 978-1-003-17386-1 (ebk)

Contents

Preface

The science of Forest Pathology is a complex amalgamation of forestry, plant pathology, ecology, and biology. The book is intended to provide comprehensive introduction to the important aspects of the field of forest pathology and tree diseases. The book is arranged in two major parts. The fundamental chapters, present forest diseases, pathogens, epidemics, and management that is applicable to all forest trees. The applied chapters on the individual crops that are grouped alphabetically present information on the symptoms, pathogen and integrated management of major diseases of forest trees. It was designed to give students a broad overview of the field of forest pathology but with sufficient detail that they will be able to assess their specific role as practicing forestry professionals. The book is intended to be the simple introductory text available. Indeed, current forestry professionals would find the text a convenient method for updating their knowledge of forest pathology. We hope that the book will prove to be a valuable tool and source of information for many years to come. It will also be useful for field workers in assessing forest tree diseases and recommending suitable management measures to conserve the forest trees and minimize the losses.

Authors

1 Introduction

Ancient Indian literature from Vedic times has numerous references to the variety of benefits that the trees and forests bestowed on mankind. Trees were worshiped. Growing trees and protecting them was considered as a good act. Forests are complex ecosystem provides a variety of valuable products like timber, fuel, fodder, minor forest produce and involving natural processes like conserving soil and water, combating desertification, maintaining biodiversity, enhancing carbon sequestration, moderating climate, homage for wildlife and providing aesthetic value and recreational needs for human. They are expected to contribute handsomely to the country's development, as evidenced in many countries. The present scenario, need is met mainly from the agroforestry sector. The forests of the naion are providing to 18% of the cattle and 16% of the human population's needs. India is also sustaining around 21.67% of the forest area under a protected area network for providing ecological security. Principal plantation species include *Acacia* spp. *Eucalyptus* spp., *Tamarindus indica*, and *Tectona grandis* are the main species established having a greater area in established forests than other species. *Eucalyptus globulus, E. grandis* and *E. tereticornis* are most common, while among the species of acacias, *Acacia auriculiformis, A. catechu, A. mearnsii, A. nilotica* and *A. tortalis* are most common. Other commonly planted broadleaf species are *Albizia* spp., *Azadirachta indica, Casuarina equisetifolia, Dalbergia sissoo, Gmelina arborea, Populus* spp. *Prosopis* spp., *Pterocarpus* spp., *Santalum album, Shorea robusta* and *Terminalia* spp. Among conifers, *Cedrus deodara* and *Pinus roxburghii* occupy a major extent; *Pinus patula* and *P. caribaea* have been planted to a limited area.

The homeland has empowered livelihood and habitation rights for forest-dependent communities along with the obligation for the conservation of forests. The strengthening in the capability of the rural poor for enlightening their income is crucial for checking the degradation of forests. The forestry sector is impacted by other sectors such as energy, agriculture, education, water resources, industry, infrastructure development, biofuels, change in demographic structure and high economic growth. It is critically important to protect these valuable resources from disturbances such as fire, pollution, invasive species, insects and diseases. To protect the forests from biotic and abiotic hazards are one of the prime objectives of forestry to sustain the environment. Forest health protection is alarmed with the conservation or development of forest and control of damage to forest by natural or man-made causes.

The commercial and recreational use of forested land is greatly affected by diseases, insects, fire, and adverse weather. Of these, diseases and wood decay cause the greatest losses and present the greatest threat to tree growth, outranking all other agents in destructiveness. The reduction of such losses is the distinct challenge and target of plant pathology. Reducing the losses of trees and other forest plants in the forest ecosystem is the special concern of the forest pathologist. Forest pathologists involve survey, conduct research on, and prescribe treatments for, the many complex diseases that affect the health of our forests. Attending to plant health issues is, of course, not new, farmers and growers have had to deal with crop losses caused by diseases or insect pests for centuries. As assert, trees have had to contend with a whole range of 'pest' threats throughout history, including insects, bacteria, viruses and fungi and various mammal pests such as deer, squirrels, rabbits, beavers and elephants. However, while dealing with tree diseases may not be an entirely new phenomenon, there is growing evidence that the incidence of an invasive tree and pathogen introductions is increasing, a trend closely linked to a signifcant upsurge in global trade and transportation in recent decades. Forest diseases happen in a difficult environmental circumstance that is a likeness of long host lifespan and fixed location. In contrast to other attentions of disease ecology, environmental influences on forest health have been accepted as important factors influencing disease in individual trees since the appearance of forest pathology as a topic of scientific inquest.

Forest Pathology

Forest pathology combines fields as diverse as botany, plant physiology, microbiology, climatology, soil science, forestry and ecology, but was born an applied science.

- » Forest pathology is the study of tree diseases. It covers major diseases of trees, considering their causes (etiology), factors that affect their spread and intensification (epidemiology), ecological and economic impacts, and management (forestpathology.org).

» Forest pathology is a remarkable field because it allows you to get into everything from molecular biology to climate change, including mycology, soils, plant anatomy and physiology, biochemistry, meteorology, silviculture, forest ecology, and more. Many forest pathologists play the roles of forest physician and conservator.

» Forest pathology is a study of tree diseases including forest trees, plantations, nurseries, urban area, and landscape settings. In addition, research of both biotic and abiotic maladies affecting the health of a forest ecosystem, primarily fungal pathogens, and their insect vectors. It is inter-discipline of forestry and plant pathology. It is a fragment of the larger approach of forest protection (Zaid Hays and Danny Watson, 2019).

» Forest pathology is the branch of botanical science / specialized branch of plant pathology that deals with the disease of forest trees for the purpose of preventing or controlling diseases. In its broadest sense, forest pathology includes diseases of trees in forest stands, diseases of shade and ornamental trees (arboriculture), and microbiological deterioration of forest products after harvest (Bier, 1950).

» Forest pathology deals with the forest trees diseases, which are mainly caused by fungal and oomycete pathogens, in both their fundamental and applied aspects. Forest pathology has obvious common roots with plant pathology, but is not simply one of its specialities (Desprez-Loustau *et al.*, 2016).

» Forest pathology is the branch of botanical science that deals with diseases of forest trees for the purpose of preventing or controlling such diseases (Boyce, 1961).

Forest Diseases

Diseases of forest trees are responsible not only for tree death, but also for reductions in growth and growth abnormalities, and can be caused either by abiotic agents, e.g. climatic and edaphic factors, chemical pollutants, or by biotic infectious agents. Diseases operate in many ways to reduce the yield and quality of timber and other values of trees. Losses begin with seed abortion and deterioration in the germinability of seeds and the survival of seedlings. Root pathogens, e.g. *Phytophthora ramorum, Fomes annosus* and *Poria weirii*, kill or decrease tree growth. Bark and cambial parasites, e.g. *Nectria galligena* and *Botryodiplodia theobromae*, can girdle and kill trees, or produce lesions that lead to decay. Wood-discoloring and rotting fungi, e.g. *Fomes pini* and *Ganoderma lucidum*, decrease the merchantability of wood in conifers and hardwoods. Foliage pathogens, such as *Phanerochaete salmonicolor, Ravenelia hobsoni* and *Elytroderma deformans* decrease the ability and quality of the economical trees and decrease the amenity value of trees, decrease their growth, and sometimes kill them.

The term "tree disease" covers a wide range of pathogenic infections, abnormalities, and disturbances of the normal structure and growth of a tree. Many different species of organisms and non-living agents can cause diseases that may reduce growth, lower quality, kill outright, or predispose a tree to attack by other pests. By definition, "tree disease" refers to the deleterious effects resulting from injurious agents other than fire, insect damage or other types of wounding (e.g. animals or hail). The causes of tree disease are usually classified as non-infectious or infectious. However, disease symptoms usually develop as a result of a complex interaction between the susceptible tree, predisposing environmental conditions, or a living, infectious agent such as a fungus. An awareness of this interaction of living and non-living agents enables greater insight into the dynamics behind disease outbreaks.

Impacts of Forest Diseases

The adverse effects of forest tree diseases may be either obvious or subtle. These are listed here without any order of relative importance:

1. **Mortality**: trees killed by any disease.
2. **Destruction of wood already formed**: meaning essentially the losses from heartrots that destroy or disintegrate wood. The decays of living trees are hence a type of loss distinct from all others.
3. **Reduction in growth increment**: decreased height and/or diameter growth which adds up to volume loss.
4. **Delayed regeneration**: years lost in getting a new crop started.
5. **Deficiencies in stocking**: too few stems, uneven spacing, or holes in the stand. Site not fully occupied and, hence, not up to full production potential.
6. **Degeneration of species composition**: disease of preferred species permitting volunteer inferior or weed species to take over the area.
7. **Deterioration of site**: soil erosion, compaction, leaching, etc.
8. **Reduction in wood quality**: incipient decay, stain, cracks, excessive knots, pitch pockets, and other responses to disease. No change in volume of production but loss in value of products.

Most of these eight types of impact are not too difficult mentally to visualize. Except for mortality, however, they are difficult ocularly to observe in the field, and are extremely difficult to measure or otherwise evaluate in quantitative terms. Also, it is obvious that any given disease may induce more than one type of loss. Reduction in growth increment eventually followed by mortality is an especially common sequence of events.

Selected References

Bakshi BK. 1976. Forest pathology – principles in practice in forestry. FRI Press, PLO, FRI, Dehra Dun., India.

Manion PD. 1991. Tree Disease Concepts. Prentice-Hall, Inc., Englewood Cliffs, NJ. 51. Manion, P. D. 2003. Evolution of concepts in forest pathology. Phytopathology 93:1052-1055.

Sinclair WA, Lyon HH. 2005. Diseases of trees and shrubs. 2nd ed. Cornell University Press, Ithaca, NY

Tainter FH, Baker FA. 1996. Principles of forest pathology. John Wiley and Sons, Inc., New York, NY.

2 Historical Aspects and Outbreaks of Forest Pathology

Prior to 1900, most tree disease research was really forest mycology. The discipline had its origins in North America about 1900, in Japan, Australia, and New Zealand in the 1930s, and in parts of Africa, South America, and the Indian subcontinent mainly since the 1950s. Most forest pathologists work in North America, Europe, and Japan, with only about 15% scattered over Asia, Australia, New Zealand, Africa, and Central and South America. Thus, the field is substantially underdeveloped in many parts of the world.

As the great Irish potato famine by late blight disease epidemics did for plant pathology, severe outbreaks caused by the introduction of exotic pathogens have marked the history of forest pathology, such as chestnut blight and white pine blister rust in North America and Europe. The identification and biological characterization of the causal agents of the diseases affecting trees were still the primary objectives of forest pathologists at the end of the nineteenth and the beginning of the twentieth century.

Pioneers in Forest Pathology

1. Robert Hartig (1839-1901)

Forest Pathology, like Forest Science, had its own origin from Europian countries. The beginning is epitomized by Robert Hartig (1839-1901), a German Forester and Professor, following in Anton de Bary's footsteps, who is considered to be the father of plant pathology. He is recognized as the father of forest pathology and founder of arboreal pathology for pioneering the field and contributing its first textbook of forest pathology in 1874. The importance of this book was recognized rapidly as is evident from its rapid translation by the famous British plant pathologist Marshall Ward and the publication of an English edition in 1894. He pronounced the Hartig net, a hyphal network of fungi that covers into the plant root. He investigated fungus hyphae in the decaying wood. He worked on important diseases of forest tree diseases caused by fungi. In 1874, he recommended the removal of infested root systems from the soil as a means of reducing future infections. Stump removal, at least in part for disease control, has been a common practice in high-value orchard crops for many years. He published the book '*Treatment of Tree diseases*' in 1882 and the *Textbook of Plant Diseases* in 1900. Along with Harry Marshall Ward and William Somerville, he published *Text-book of the Diseases of Trees.*

2. George Henry Hepting (1907-1974)

George Henry Hepting is an America's scientist, a specialist in the theory and practice of forest pathology. He devoted his remarkably energetic life to learning, understanding, and teaching how trees survive disease stresses induced by biotic and abiotic agents in forests. His significant researches include fire scars, basal wounds, stump sprouts, little-leaf disease of southern pines, wilt disease of mimosa, pitch canker disease of southern pines, oak wilt, white pine blister rust, and fusiform rust. His 1971 text, *Diseases of Forest and Shade Trees in the United States* provides the most comprehensive encyclopedia of knowledge on these topics. He wrote a definitive history of the failure of efforts to control chestnut blight and similarly ineffective attempts to control the Dutch elm disease after these diseases were introduced into North America. Hepting also had an influential role in the increased participation of forest pathologists in international forestry policy discussions and in the activities of the American Phytopathological Society, including the establishment of the subject matter committee on Forest Pathology. He was an associate editor of *Phytopathology*, and for a number of years was a member of the Editorial Board of the *Annual Review of Phytopathology.* Hepting's achievements in forest pathology were recognized by many honors and awards.

3. Paul D Manion (1981-2003)

Paul D Manion formalized the "tree decline concept," with biotic and abiotic factors acting and interacting in sequence as predisposing, inciting, or contributing factors. He published few texts *Tree Disease Concepts, Evolution of Concepts in Forest Pathology* and *Forest Decline Concepts.*

Beginners of Forest Pathology in India

1. Leslie Charles Coleman

Leslie Charles Coleman (1878-1954) was a Canadian plant pathologist, virologist and entomologist who served as the first director of agriculture in Mysore State in southern India. He conducted pioneering research on the pests and diseases affecting agriculture in the region. His major work to plant pathology included methods to control a kolerago rot of coffee caused by *Pellicularia koleroga.* Sprays of inexpensive Bordeaux mixture on the growing crowns helped control infection caused by what he described as *Phytophthora arecae.* Coleman established research on sandal spike disease in Karnataka.

2. Dr. Bimal Kumar Bakshi

Bimal Kumar Bakshi is an eminent forest pathologist and served as a Director, Forest Research Institute, Dehradun, India. He published many articles and books related to forest diseases in India. His publications includes *Fungi associated with ambrosia beetles in Great Britain, Indian Polyporaceae* (on trees and timber), and *Forest Pathology: Principles and Practice in Forestry.*

Historical Outbreak of Tree Diseases Around World

Outbreaks of tree disease with destructive effects are associated either with the emergence of native pathogens due, for instance, to intensive forest management or fire suppression or can occur when pathogen or host introductions lead to novel host–pathogen interactions. Tremendous epidemics have been attributed to the lack of co-evolution between hosts and pathogens. Climate change may also be responsible for an increase of outbreaks, especially of necrotrophic pathogens. Diseases affected a relatively small extent of forests in all geographic regions. Asia reported over 5 million hectares of forest affected by diseases, and Europe reported just less than 5 million hectares affected (van Lierop, 2015). The major outbreak of infectious tree diseases in the world are listed below

1. Dutch elm disease

Dutch elm disease is one of the world's most serious tree diseases. It is caused by the fungus *Ophiostoma novo-ulmi*, which invades and blocks the water conducting systems of trees. This results in the wilting and death of the tree. Dutch elm disease first appeared in the north-west of Europe about 1910. Between 1914 and 1919, several Dutch scientists carried out influential research on the cause of the disease. Then in the late 1920s UK Forestry Commission began to monitor its rapid spread to Britain. In 1921 it was first repoted in

Holland. The first epidemic caused losses of 10-40% of elms in several European countries. Within an era about 30 million elms were dead. By the 2010 it was considered some 60 million elm trees have been lost to the disease. Studies on the new Dutch elm disease fungus showed that it contrasted from the original fungus in almost all its important pathological properties. The two pathogens were later described as separate species:

a. *Ophiostoma ulmi* - caused the original epidemic

b. *Ophiostoma novo* - a new highly aggressive pathogen that caused the second epidemic.

2. Chestnut blight

Sweet chestnut blight is a vicious disease of sweet chestnut trees caused by the ascomycete fungus *Cryphonectria parasitica.* It was formerly known as *Endothia parasitica.* It can also affect a small number of other plant species. Sweet chestnut blight occurs in Japan, China, Korea, USA, Canada, Italy and throughout Europe and infects chestnut, oak, red maple, shagbarkhickory, and eucalyptus. Origially detected in 1904 at the New York Zoolagical Park, the disease spread within a span of 50 years to virtually eliminate the chestnut throughout its botanical range. It is caused by the fungus *Cryphonectria parasitica* and infects American chestnut trees (*Castanea dentata*) throughout the United States and Canada in early 1900s. The fungus arrived from Asia (imported from Japanese chestnut trees) in the late 19th century. By 1913, the disease had wiped out enough trees to warrant investigation by the United States Department of Agriculture. By 1940, over 3.5 billion trees had been lost to the disease. *Cryphonectria parasitica* infection is usually fatal to European and North American sweet chestnut trees. It distressed sweet chestnut forests in eastern United States during the beginning of the 20th century, killing an estimated 3.5 billion trees after it was inadvertently introduced from Asia, probably during the 19th century.

3. Beech bark disease

Beech bark disease refers to an insect-fungal complex that consists of a sap-feeding scale insect and at least two species of *Nectria* fungi. Beech bark disease begins when American beech (*Fagus grandifolia*) becomes infested with beech scale (*Cryptococcus fagisuga*) (=*Cryptococcus fagi*) during 1920s - 1930s. Minute wounds and injuries initiated by the scale insects eventually support the *Nectria* fungus to enter the tree. *Nectria* kills parts of woody tissue, sometimes creating cankers on the tree stem and large branches. If enough tissue is killed, the tree will be girdled and die. Beech scale was accidentally introduced into Nova Scotia in 1890 on ornamental beech imported from Europe. By the early 1930s, the scale and associated *Nectria* fungi were found throughout the Maritime Provinces in eastern Canada and in localized areas of Maine. Much of New England and parts of New York were affected by 1960, and the disease had extended into northeastern Pennsylvania by 1975. Beech bark disease has continual to spread throughout Northern America.

4. Butternut canker

Butternut canker disease disturbs butternut trees, and ensues when the *Sirococcus clavigignenti-juglandacearum* fungus enters the tree and causes cankers to form on the trunk, branches, or exposed roots. Canker development on the trunk can girdle the tree, leading to canopy dieback and eventual tree mortality. Butternut canker disease is resent throughout the natural range of the butternut tree in North America, which is especially worrisome as this tree is endangered in eastern Canada. The origin of the fungus that causes butternut canker disease is not known, though some scientists trace it back to Asia.

5. Sudden oak death

Sudden oak death caused by the oomycetes *Phytophthora ramorum.* It causes rapid decline and mortality of Japanese larch in European conifer plantations. The first symptoms may be crown dieback and premature browning of infected needles. Bleeding cankers are found on branches and main stems, most often located at the top of the tree, and lead to eventual death. The disease kills oak and other related tree species and had devastating effects on the oak populations in Oregon and California, as well as being present in Europe.

6. White pine blister rust

White pine blister rust is probably the most destructive disease of five-needle (white) pines in North America caused by *Cronartium ribicola.* It was detected in 1854 in the Baltic provinces of Russia followed by 1906 at Geneva, New York and 1922 at Europe. A small, yellow or red spot appears on the needle at the site of infection and the needle may die as the fungus grows into the bark. Damage shows up as dead or dying branches. The pathogen initiated in Asia and became established in Europe in the 18th century after highly susceptible American white pines were widely planted. The pathogen was entered into North America about 1900s on white pine seedlings grown in European nurseries and by the 1950s had spread to most of the commercial white pine regions in the world.

7. Jarrah die-back

Jarrah is *Eucalyptus marginata,* as a valuable timber tree of native forests and Southern in

Australia. Canker disease causing die-back of the trees by the fungus. *Phytophthora cinnamomi* was first described in 1922 cinnamon plants in Sumatra, Indonesia. The disease was recently into the forests of Southern Australia. Jarrah trees an under storey plants in South-western Australia forests were killed by this disease. They were replaced by grasses and sedges. The same pathogen is also responsible for the death of mangroves in Queensland. This epidemic disease has disrupted ecology and the ecosystem.

8. Fire blight of pome

Fire blight bacterium, *Erwinia amylovora* attacks many rosaceous plants including fruit trees like pear and apple. Fire bight organ bacterium was first observed in New York in 1780. It was introduced in United Kingdom in 1957 probably from New Zealand, in 1966 in the Netherlands and 1978 in France. The disease kills numerous pear and apple trees every year. It is the cause for yield decline in pear in the United States and Europe. In Italy it has eradicated 500,000 Pear trees since 1995. Severe losses were experienced in apple and tea France in 1984 and in Switzerland in 1995. In Crimea, in Pear varieties, the bacterium caused a loss of 60-90% of flowers and buds. Outbreak in Egypt in 1982-1984 was due to flower blight and it caused loss of 10-75% flowers/trees. An epidemic fire blight in South-west Michigan apple orchards in 2000 was severe. It was estimated that 3,50,000 - 4,50,000 apple trees were killed and 1500-2300 acres of apple orchards were lost and the developmental costs of these orchards was the US $ 9 million. In this area apple yields reduced on an average 35% and some growers lost 100%. Economic loss in this region was estimated at $ 42 million in 2000.

9. Pine wilt

Pine wilt is caused by pine wood nematode *Bursaphelenchus xylophilus*. This pine wilt disease, which is endemic to North America, has already spread epidemically to East Asia, and recently to Portugal in Europe. *Monochamus* spp. beetles (pine sawyer beetles) are the only insects that disperse the pine wood nematode. It was initially recognized as unusual wilt of pines in 1905 in the port city of Nagasaki in southern Japan, which was the only port open to foreign culture in the nineteenth century. The loss of timber by pine wilt showed an abnormal increase in the late 1940s soon after World War II and then pine wilt became a matter of public concern.

Historical Outbreak of Tree Diseases in India

1. Shisham mortality

Dalbergia sissoo (Shisham) is a common tree species for the manufacture of quality timber in India. During the 1960s states like Bihar, Haryana, Delhi, Punjab, Himachal Pradesh and, Uttar Pradesh, sissoo mortality was observed to be prominent either in isolated trees or on the plants growing on field bunds, road sides and canal sides. In the visible roots of the stressed trees it was observed that the feeding roots were blackish in colour and partially decomposed. In few cases, block plantations as well as diversified cropping, exhibited fewer mortality. However, the trees growing on the fringes are being affected. An ocular estimation of the area has revealed about 20-30% mortality in India. The problem is regional in nature and stands of this species have been affected in Bangladesh, India, Nepal, and Pakistan. The resulting diseases were described to be accompanying with sissoo mortality.

a) *Fusarium* wilt

The *Fusarium* wilt disease caused by *Fusarium solani* f.sp. *dalbergiae* has been reported from Uttar Pradesh, Bihar and Punjab in plantations, raised on unsuitable sites i.e., stiff, clayey soils and water-logged conditions. It does not occur in normal forests and in plantations developed on suitable sites. It causes 60-80 % losses particularly in heavy and wet soils (Kaushik and Singh, 1996).

The characteristic symptoms of the disease are yellowing and deaths of leaves in acropetal succession up the tree. Eventually, the entire tree appears chlorotic. In progressive stages of the disease, the leaves are shed rendering the branches plain. The infected trees die within a few days or months. Trees of advanced age are usually susceptible to the disease. The outer sapwood parades a distinctive pink to reddish stain. Though the stain is constrained in this region, it may seldom penetrate into the inner sapwood. The heartwood is free from the stain. The stain also evolvements alongwith the outer sapwood from the roots to the stem and, in the late stage of wilt; it may extend up the stem to about 3m from the ground (Bakshi 1954). The pathogen is mostly constrained to roots. The hyphae and jelly-like substances plug the vascular tissues resulting in wilt symptoms. Unrevealed toxins are synthesized in the culture filtrate, which may also be responsible for causing wilt (Bakshi and Singh 1959). Eradication of the pathogen from the soil is not possible either chemically or by crop rotation (Bakshi 1955). It causes 60-80 % losses particularly in heavy and wet soils. Proper site selection with light textured soil, adequate soil moisture and good drainage is important for raising healthy plantations free from wilt disease.

b) *Ganoderma* root rot

Sissoo root rot was due to *Ganoderma lucidum* in some parts of India. It is common both in natural forests and in plantations. However, a solemn disease problem may arise when Sissoo is grown as a reforested pure crop without eradicating the stumps and roots of the original crop. The disease is prevalent in both light and heavy textured soils. In light soils, the pathogen spreads rapidly on roots and the trees are killed in a short period. However, in rigid soils, the fungus spreads gradually on roots. The pathogen is a root inhabitant and infects the roots through intact as well as injured surfaces. It causes white spongy rot in the sapwood. The affected trees show a stag-headed look and are ultimately killed. The symptoms of the disease appear when a large portion of the root system is injured by the pathogen. Lateral spread of the disease in plantations is through root contact, which results in gaps in wholesome plantations.

c) *Phellinus* root and butt rot

Root and butt rot of Sissoo caused by *Phellinus gilvus*. It has been studied by Bakshi (1971). The fungus is primarily a wound parasite and is known to infect plantation trees. It occurs in association with *Ganoderma lucidum* or the wilt pathogen, *Fusarium solani*. Trees of progressive age are criticized and they exhibit stag-headed conditions. The fungus causes controlled decay as white rot in the sapwood and to a restricted extent in the heartwood. Sporophores of the fungus are established on root and stem bases. They are annual, sessile, usually reflexed, rarely effuse-reflexed, leathery when fresh, drying hard, single or imbricate; upper surface with shades of yellow, brown and red and either coarsely hairy, sub-zonate in forma gilvoides or smooth with concentric zonations in forma licnoides; lower surface yellowish-brown, bores round, dissepiments thick, margin, sterile (Bakshi 1971, 1976).

d) Root-knot nematode

Meloidogyne javanica has been recorded to form galls on roots of sissoo from Lachhiwala range nursery, Dehra Dun (Mehrotra and Sharma 1992). The nematode is polyphagus and has been found to attack several other tree species (Sharma and Mehrotra 1992). Heavy infestations of the nematode affect plant growth adversely.

2. *Scleroderris* canker

Scleroderris canker or *Brunchorstia* disease has been a worrisome and stressful problem across Canada in the early to mid-1980s. It is caused by a fungal pathogen *Gremmeniella abietina*, infecting coniferous forests. The main symptom is the decease of the needles, leading to the demise of the tree. This disease was already causing much damage in young pine plantations in eastern Canada when a more destructive race (called the European race) of the causal fungus was discovered in New York State in 1976. It was infesting and even killing mature pine stands within 3-4 years after infection.

3. Sandal spike disease

Sandal (*Santalum album* L.) is a commercially important, perennial, semi root parasite, evergreen tree species. The sandal spike disease is becoming a serious threat to sandal cultivation in the southern states *viz.*, Karnataka, Kerala, Madhya Pradesh, Maharashtra, Andra Pradesh and Tamil Nadu. The disease is categorized by witches'-broom symptoms, containing of small, narrow leaves that turn pale-green or yellow on branches. Infected leaves become overcrowded due to internodes shortening and standout stiffly from branches, acquiring a spike-like appearance Leaves and branches dry out and affected plants eventually die-off within a couple of years from the first appearance of symptoms. Research to investigate the causal organism for sandal spike disease spanned over more than half a century, from 1900 to 1968, but without any conclusive findings (Barber 1903; Ayyar 1918; Hole 1918; Coleman 1923; Sreenivasaya 1930; Ramaiah et al. 1962; Parthasarathi et al. 1966). Dijkstra and Ie (1969) observed the presence of mycoplasma in the phloem sap of infected S. album trees, with Hull et al. (1969) and Verma et al. (1969) independently confirming the causal agent to be a mycoplasma. The average annual production of sandal wood has been drastically declined from 4000 tonnes to 1044 tonnes in 1965 to 1995 (Srinivasan et al., 1992). The incidence of Sandal spike disease in southern Indian states has become a predominant constraint for the successful management of sandal plantations.

4. Reed bamboos witches' broom

A disease witches' broom caused by the *Balansia linearis* fungus affecting reed bamboos was recorded during a disease survey in natural stands and plantations in Kerala State, India. Depending on host species and site, the pathogen infected up to 24% of the clumps of the commercially exploited reed bamboos, *Ochlandra travancorica* var. *hirsuta*, *O. travancorica*, *O. scriptoria*, and *O. ebracteata* poses a menace to the reed bamboo cultivation (Mohanan, 2004).

5. *Casuarina* wilt

Wilt or blister bark disease is the most destructive disease of *Casuarina equisitfolia* in India. It is caused by the fungus, *Subramanianospora vesiculosa* (Narayanan, Sharma and Minter, 2003), and was first reported from India in the early 1900s. Later outbreaks were reported from India, Mauritius and Sri Lanka (Narayanan et al., 1996). In recent years the disease has been reported to cause large-scale tree mortality in India and other countries (Narayanan, Sharma and Minter, 2003). Discoloration of the leaves is the preliminary symptom of blister bark disease. As the disease progresses, necrotic lesions appear all over the main stem and branches. Subsequently, all the affected trees reveal symptoms of wilting and drying and are ultimately killed. Likely pathways of introduction and spread include nursery stock and bark and wood packaging material, including dunnage.

Selected References

Bagchee KD. 1945. Pathological notes No. 2 - Wilt and dieback of Shisham, Babul and Khair in the artificial regeneration under agriculture-cum-forestry management. Indian Forester, 71(1), 20-24.

Bakshi BK, Arora KK, Singh S. 1957. Root disease of shisham *(Dalbergia sissoo)*. V. Incidence of disease in relation to soil pH and soil texture. Indian Forester ,83(9), 555-558.

Bakshi BK, Singh SL. 1959. Root disease of shisham *(Dalbergia sissoo)*. VIII Inoculation studies on wilt. Indian Forester, 85(7), 415-421.

Bakshi BK. 1954. Wilt of shisham *(Dalbergia sissoo* Roxb) due to *Fusarium solani* Sensu Snyder and Hansen. Nature, 174, 278-91.

Bakshi BK. 1955. Wilt disease of shisham (*Dalbergia sissoo* Roxb). II. Behaviour of *Fusarium solani*, the wilt organism in soil. Indian Forester 81(4), 276-281.

Bakshi BK. 1957. Wilt disease of shisham (*Dalbergia sissoo* Roxb). IV. The effect of soil moisture on the growth and survival of *Fusarium solani* in laboratory. Indian Forester 83(8), 505-512.

Bakshi BK. 1974. Control of root disease in plantation in reforested stands (with special reference to Khair, Sissoo, Eucalyptus etc.). Indian Forester 100(1), 77-78.

Bakshi BK. 1976. Forest Pathology – Principles in practice in Forestry. FRI Press, PLO, FRI, Dehra Dun., India.

Barber CA. 1903. Report on spike disease in sandalwood trees in Coorg. Indian Forester 29, 22–31.

Boyce JS, Bakshi BK. 1959. Dying of sal. Indian Forester, 85, 585-588.

Champion HG, Seth SK. 1968. A revised survey of forest types of India. Manager of publications Delhi, Govt. of India Press.

Desprez-Loustau M, Aguayo J, Dutech C. 2016. An evolutionary ecology perspective to address forest pathology challenges of today and tomorrow. Annals of Forest Science 73, 45–67. https://doi.org/10.1007/s13595-015-0487-4.

Hole RS. 1917. Cause of the spike disease of sandal (*Santalum album*). Indian Forester 40, 430–442.

Hole RS. 1918. Spike disease of sandal. Indian Forester, 44, 325–334.

Houston DR, O'Brien JT. 1983. Beech bark disease. Forest Insect and Disease Leaflet No. 75. USDA Forest Service, Northeastern Forest Experiment Station, Hamden, Conn. Houston, D.R., and H.T. Valentine. 1988. Beech bark disease: the temporal pattern of cankering in aftermath forests of Maine. Canadian Journal of Forest Research 18:38-42.

Kaushik JC, Singh A. 1996. Management of seedling root rot of *Dalbergia sissoo* caused by *Fusarium solani*. Proceedings, IUFRO Symposium on Impact of Diseases and Insect Pests in Tropical Forests. pp. 209-216.

Maloy OC. 1997. White pine blister rust control in North America: A case history. Annu. Rev. Phytopath. 35, 87-109.

Manion P. 1988. Hardwood forest declines—concepts and management. Pages 127-130 in: Proc. Soc. Am. For. Natl. Conv. Rochester, NY. 50.

Manion PD, Lachance D. eds. 1992. Forest Decline Concepts. American Phytopathological Society, St. Paul, MN. 54.

Manion PD, Lachance, D. 1992. Forest decline concepts: An overview. Pages 181-190 in Forest Decline Concepts. P. D. Manion and D. Lachance, eds. American Phytopathological Society, St. Paul, MN.

Manion PD. 1991. Tree Disease Concepts. Prentice-Hall, Inc., Englewood Cliffs, NJ. 51. Manion PD. 2003. Evolution of concepts in forest pathology. Phytopathology 93:1052-1055.

Mehrotra MD, Sharama V. 1992. Some new host record of root knot nematodes. Indian Forester, 118, 856-857.

Mohanan C. 2004. Witch's broom disease of reed bamboos in Kerala, India. Forest Pathology, 34(5), 329.

Murali R, Rangaswamy KT. Nagaraja H. 2019. Occurrence of spike disease in sandal plantations in southern karnataka and cross transmission studies between sandal spike and stachytarpheta phyllody through dodder (*Cuscuta subinclusa*). Int. J. Curr. Microbiol. App. Sci. 8(01), 1041-1046. doi: https://doi.org/10.20546/ijcmas.2019.801.113

Narayanan C, Sharma JK, Minter DW. 2003. *Subramanianospora vesiculosa* - a hyphomycete causing wilt disease of *Casuarina equisetifolia*. Indian Phytopathology, 56(2), 159-163.

Negi JDS, Pokhriyal TC, Sharma SD, Bhandari RS. 1999. Shisham mortality in India – A case study. Unpublished report.

Parthasarathi K, Gupta SK, Rao PS. 1966. Studies on Sandal Spike. Part VII. Some useful stain tests for the diagnosis of the spike disease. Proc. Indian Acad. Sci. Sect. B 64, 152–156.

Ramaiah PK, Parthasarathi K, Rao PS. 1962. Studies on sandal spike. IV. A histochemical study of the sandal (*Santalum album* Linn.) root haustoria in relation to mineral nutrition. Proc. Indian Acad. Sci. Sect. B 56, 111–113.

Sharama V, Sharama V. 1992. *Meloidogyna* species cause of root knot of important forestry tree species in nursery. Indian Forester 118, 961-962.

Sinclair WA, Lyon HH. 2005. Diseases of trees and shrubs. 2nd ed. Cornell University Press, Ithaca, NY

Tainter FH, Baker FA. 1996. Principles of forest pathology. John Wiley and Sons, Inc., New York, NY.

van Lierop P, Lindquist E, Sathyapala S, Franceschini G. 2015 Global forest area disturbance from fire, insect pests, diseases and severe weather events. For. Ecol. Manage 352, 78–88.

3 Plant Disease: Symptoms and Signs

Plant diseases are recognized by the symptoms (external or internal) produced by them or by sick appearance of the plant. The term plant disease signifies the condition of the plant due to disease or cause of the disease. Plant disease is mainly defined in terms of the damage caused to the plant or to its organ. The other definitions for the term disease are:

» Disease is a malfunctioning process that is caused by continuous irritation, which results in some suffering producing symptoms. This definition is accepted by both American Phytopathological Society and British Mycological Society.

» Disease is an alteration in one or more of the ordered sequential series of physiological processes culminating in a loss of coordination of energy utilization in a plant as a result of the continuous irritation from the presence or absence of some factor or agent.

» A plant is said to be 'diseased' when there is a harmful deviation from normal functioning of physiological process (Federation of British Plant Pathologists, 1973).

» The disease can also be defined as 'any disturbance brought about by a living entity or non-living agents or environmental factors which interfere with manufacture, translocation or utilization of food, mineral nutrients and water in such a way that the affected plant changes in appearance with or without much loss in yield than that of a normal healthy plant of the same variety.

Symptomatology, which deals with (a) the presence of readily apparent morphological abnormalities of the host in response to a causal agent (symptoms) and (b) an indication of disease other than a reaction of the host, such as the mycelium or fruiting body of a causal fungus (sign).

Symptoms: any alterations in the appearance of the host due to disease or visible or detectable abnormality in a plant as a result of a disease or disorder. It is the manifestation of a disease. It is due to the morphological (external) and physiological (internal) reactions or alterations of a disease.

a. **Morphological symptoms:** Symptoms appear externally on the whole part of diseased plant. They are detected by visual means

b. **Physiological or histological symptoms:** Symptoms appear in the tissues of affected plants which are detected only by microscopic examination of diseased tissue.

Examples:

» Root rots: crown thinning, dieback, decay, resinosis etc.

» Foliage diseases: discolorations, lesions, defoliation etc.

Symptoms are further divided into three general categories:

1. Necrotic symptoms, or those symptoms that result from the cessation of function leading to death;
2. Hypoplastic symptoms, or those symptoms that result from underdevelopment or retardation of function;
3. Hyperplastic symptoms, or those symptoms that result from over development or acceleration of function.

Necrotic symptoms

1. **Blight**: the rapid killing of foliage, blossoms, and twigs.
2. **Blotch**: large, irregular lesions on leaves, shoot, and stems.
3. **Canker**: necrotic, often sunken lesions in the cortical tissues of stems and roots, death of cambium.
4. **Decay**: the disintegration of dead tissues, death of paranchyma cells in wood of roots and stem of living trees.

5. **Decline**: defoliation occurs due to abiotic stress and pathogen infection.
6. **Dieback**: progressive death of twigs and branches from the tips toward the trunk, sudden dying of all or part of tree due to wilt.
7. **Hydrosis**: a water-soaked, translucent condition of the tissue due to cell sap passing into intercellular spaces.
8. **Scald**: blanching of the epidermis and adjacent tissues.
9. **Scorch**: browning of leaf margins resulting from the death of tissues.
10. **Shot hole**: circular holes in leaves resulting from the dropping out of the central necrotic areas of spots.
11. **Spot**: lesion, usually defined, circular, or oval in shape, with a central necrotic area surrounded by variously colored zones.
12. **Wilt**: leaves or shoots lose their turgidity and droop.
13. **Yellowing**: leaves turn yellow due to the degeneration of chlorophyll.

Hypoplastic symptoms

1. **Chlorosis:** failure of chlorophyll development in normally green tissues.
2. **Dwarfing:** subnormal size in an entire plant or some of its parts.
3. **Etiolation**: yellowing due to lack of light.
4. **Rosetting**: a crowded condition of foliage due to lack of internode elongation.
5. **Suppression**: prevention of the development of certain organs.

Hyperplastic symptoms

1. **Anthocyanescence**: purplish or reddish coloration of leaves or other organs due to overdevelopment of anthocyanin pigment.
2. **Callus**: overgrowth of tissues at the margins of wounds and diseased tissues.
3. **Curl**: rolling or folding of leaves due to localized overgrowths of tissue.
4. **Fasciation**: a flattened condition of a plant part that is normally round.

5. **Fasciculation or witches'-broom**: broomlike growth of densely clustered branches.
6. **Gummosis**: excessive production of gums at the margins of wounds and on diseased tissues.
7. **Heterotopy**: development of more or less normal tissues or organs in the wrong place.
8. **Resinosis**: excessive production of resin flow at the margins of wounds and on diseased tissues.
9. **Sarcody**: an abnormal swelling of tissues above girdled branches or stems.
10. **Scab**: roughened, crustlike lesion.
11. **Tumefaction**: tumorlike or gall-like overgrowth of tissue.
12. **Virescence**: development of chlorophyll in tissues where it is normally absent.
13. **Witches broom**: Proliferation of adventious buds.
14. **Leaf blister**: localized enlargements causing puckering.

Signs

Signs: the physical appearance of pathogen or the visible structure of the pathogen produced in or on the diseased tissues.

1. **Mycelium**: filamentous mycelium or mold growth, under some conditions, readily visible to the naked eye.
2. **Conks and mushrooms**: the familiar structures of some fungi that are formed by some pathogenic fungi.
3. **Fruiting bodies**: reproductive structures of some fungi that are embedded in diseased tissue, often requiring a hand lens to visialize the fruiting bodies.
4. **Sclerotia**: hard resistant structures in root rots.
5. **Rust pustules**: Orange yellow or reddish colour erumpent raised basiocarp growth on the surface.
6. **Bacterial exudate**: ooze and specific odors associated with tissue macerations by certain pathogens.

7. **Superficial mycelial growth**: white powdery growth on leaf in powdery and downy mildew diseases.

Examples:

» White trunk rot: conk of *Phellinus igniarius*

» *Armillaria* root rot: mushroom, rhizomorph, mycelial fan of fungi

» Laminated root rot: conk of *Phellinus weirii*, setal hyphae

Signs are further divided into three general categories: (1) vegetative structures, or those pathogen structures that function primarily in absorption and storage of nutrients; (2) reproductive structures, or those pathogen structures that function in reproduction of the organism; and (3) disease products, or gases and exudation products resulting from disease.

Vegetative structures

1. **Felt**: a densely woven mat of mycelium
2. **Haustorium**: absorbing organ of a fungus that penetrates a host cell without penetrating the plasma membrane
3. **Mycelium**: a mass of fungal threads or hyphae
4. **Pathogen cells**: generally, masses of bacterial cells
5. **Plasmodium**: a naked mass of protoplasm
6. **Rhizomorph**: a compact, macroscopic, cordlike strand of fungus tissue resembling a root
7. **Sclerotium**: hard, compact, resting body composed of fungal hyphae

Reproductive structures

1. **Acervulus**: a subcuticular or subepidermal mat of hyphae, conidiophores, and conidia that are not covered by fungal tissue and are generally associated with a host
2. **Apothecium**: open, cuplike, ascus-containing fruiting body
3. **Ascus**: saclike structure containing ascospores formed as a result of karyogamy and

meiosis

4. **Basidium**: characteristically club-shaped structure on which basidiospores are produced as a result of karyogamy and meiosis

5. **Cleistothecium**: closed ascus-containing fruiting body

6. **Conidiophore**: specialized hyphal branch on which conidia are produced

7. **Conk**: woody shelflike structure characteristic of many wood-rotting fungi

8. **Mildew**: cobwebby or powdery superficial growth, usually on leaves

9. **Mold**: wooly or furry surface growth of mycelium, conidiophores, and conidia

10. **Mushroom (=toadstool)**: umbrella-shaped fruiting structure of many basidiomycetes

11. **Perithecium**: characteristically flask-shaped, ascus-containing fruiting body with a small opening (ostiole) and a wall of its own

12. **Pseudothecium (=ascostroma)**: fruiting body bearing asci in locules within a stroma

13. **Pycnidium**: asexual sporocarp containing conidia in a cavity

14. **Seed-bearing plants**: higher plants that parasitize trees

15. **Sorus**: mass or cluster of spores borne on short stalks

16. **Sporangium**: the enlarged tip of specialized hyphal branch in which sporangiospores are produced

17. **Spore**: general name for a single- to several-celled propagative unit in fungi and lower plants. Examples of spores with specific names are conidia, ascospores, basidiospores, zoospores, oospores, sporangiospores, aeciospores, urediospores, chlamydospores, and teliospores

18. **Sporodochium**: cushion-shaped stroma covered with conidiophores

19. **Stroma**: the compact mass of fungal hyphae on or within which fruiting structures are formed.

20. **Worms**: generally nematodes, which are microscopic, wormlike animals that can cause disease.

Classification of Diseases

- By maturity or management scheme:, and those of mature trees.

Classification of forest infectious diseases

1. Based on the cause of diseases

- Non-infectious or non-parasitic or non-pathogenic or abiotic diseases.
- Infectious or parasitic or pathogenic or biotic diseases.

2. Based on the maturity or management of diseases

- Nursery diseases
- Sapling diseases
- Plantation diseases

3. Based on the severity of diseases

- Decline complexes (Dieback)
- Vascular wilt
- Foliage diseases
- Root and butt rots
- Stem and Broom rusts
- Stem decay and cankers

4. Based on type of trees

a. Conifer diseases

- Bud blights
- Canker
- Die-back
- Felt blights and snow molds

- Mycelial fan
- Needle blight
- Needle cast
- Punk knot
- Resinosis
- Root rot
- Rust
- Witches broom

b. Broad leaf disease/ hardwood disease

- Anthracnose
- Bracket
- Butt rot
- Chlorosis
- Conk
- Galls
- Leaf scorch
- Leaf spot
- Powdery mildew
- Root rot
- Rust
- Tar spot
- Wood decay

5. Based on the part of trees

a. Root diseases

» Butt rot

» Cortical root rot

» Heart rot

» Mycelial fan

» Resinosis

» Root decay

» Vascular wilt

b. Stem (Phloem and cambium) diseases

» Alder heat canker

» Aspen cankers

» Bacterial cankers

» Beech bark disease

» Blisters

» Bracket

» Butternut canker

» Canker

» Canker

» Chestnut blight

» Conk

» *Cytospora* cankers

» Fasciation

- Galls
- Heart decay
- Parasitic plants
- Pitch canker
- Punk knot
- Rust
- Salt-and-Pepper (*Hypoxylon*) canker
- Shoot blight
- Some target cankers (Annual canker, Perennial canker, Diffuse canker)
- Sooty-bark disease of maple
- Splits
- Stem decay
- Sudden oak death
- Thousand cankers disease
- Water and gummy exudations
- Wood decay

c. Foliar diseases

- Anthracnose
- Anthracnose
- Bacterial Leaf Scorch
- Bronzing
- Browning
- Chlorosis

- Color changes
- Crown Gall
- Die-back
- Drippy nut
- Leaf blight
- Leaf distortion and reduction
- Leaf scorch
- Leaf spot
- Needle blight
- Needle cast
- Needle cast
- Pod rot
- Powdery mildew
- Rust
- Sooty mould
- Tar spot
- Wetwood
- Witches broom
- Yellowing

6. Based on pathogens

a. Oomycete disease

- **Daming off** - *Pythium* sp., *Phytophthora* sp.
- **Die-back**- *Phytophthora* sp.
- **Leaf blight**- *Phytophthora* sp.

» **Root rot-** *Phytophthora* sp.

b. Fungal disease

» **Canker-** *Diaporthe lokoyae*

» **Cone, seed and seedling disease-** *Sirococcus strobilinus*

» **Daming off-** *Rhizoctonia* sp., *Phoma* sp.

» **Galls-** *Endocronartium harknessi*

» **Leaf and needle rusts-** *Melampsora* sp. and *Pucciniastrum* sp.

» **Needle cast-** *Lophodermella concolor*

» **Root rot-** *Fusarium* sp., *Phoma* sp. and *Cylindrocladium* sp.

» **Stump infection-** *Heterobasidion annosum*

» **Wood decay-** *Armillaria* sp.

c. Bacterial disease

» **Leaf blight-** *Xanthomonas* sp.

» **Shot hole and fruit canker-** *Xanthomonas* sp.

» **Wilt-** *Ralstonia solanacearum*

d. Phytoplasmal (bacteria) disease

» **Little leaf-** *Candidatus* Phytoplasma

» **Spike-** *Candidatus* Phytoplasma

» **Witches' broom-** *Candidatus* Phytoplasma

» **Yellow dwarf-** *Candidatus* Phytoplasma

e. Viral disease

» **Bushy stunt-** *Scots pine bushy stunt virus*

» **Mosaic-** *Bamboo mosaic virus, Poplar mosaic virus*

» **Mottle-** *Elm mottle virus*

» **Ring spot-** *Prunus necrotic ringspot virus*

f. Parasitic plants

» **Dwarf mistletoe-** *Arceuthobium* sp.

» **Gaint mistletoe-** *Dendrophthoe* sp.

» **Leafy mistletoes-** *Phoradendron* sp.

» **True mistletoes-** *Viscum* sp.

7. Based on the perpetuation and spread

» Air borne disease

» Seed/ seedling borne disease

» Soil borne disease

8. Based on occurrence and geographic distribution

» Endemic disease

» Epidemic disease

» Pandamic disease

» Sporadic disease

Plant Disease Detection and Diagnosis

Greater understanding of forest disease symptoms often involves much careful study. Complicating factors such as predisposing weather, secondary fungi and genetic variation in a single species of fungus or host, require that each step of an investigation be carefully undertaken according to scientific principles. The following aspects of a pathogen and techniques used in its study are important to forest pathology diagnosis:

» **Culturing** - the suspected pathogen should be obtained in a pure, artificial culture, where possible.

» **Pathogenicity study** - according to Kochs' Postulates, the organism should be proven to be the cause of an infectious disease.

» **Life cycle** - all spore states should be identified and studied and related to host range and phenology.

» **Conditions for infection** - the general physiology and requirements for spore germination, penetration, and so on should be elucidated.

Koch's Postulates

1. Show constant association of the organism with the disease.
2. Isolate the organism in culture from the diseased plant.
3. Inoculate a healthy plant from the culture and produce the same disease.
4. Re-isolate the same organism from inoculated plants.

Selected References

Andrivon D. 1995. Nomenclature for pathogenicity and virulence: Precision vs. tradition. Phytopathology 85, 518-519.

Browne FG. 1968. Pests and Diseases of Forest Plantation Trees. Clarendon Press, Oxford, UK.

Buckland DC, Redmond DR, Pomerleau R. 1957. Definition of terms in forest and shade tree disease. Can J. Bot. 35, 675-679.

Büttner C, von Bargen S, Bandte M, Mühlbach HP. 2013. Forest diseases caused by viruses. In: Infectious forest diseases. Gonthier P., Nicolotti G. (eds), CABI, p. 50-75.

Ehrlich J. 1941. Etiological terminology. Chron. Bot. 6, 248-249.

Hawksworth DL, Kirk PM, Sutton BC, Pegler DN. 1995. "Ainsworth and Bisby's Dictionary of the Fungi," 8th Ed. Commonwealth Mycological Institute, Kew, Surrey.

Hepting GH, Cowling EB. 1977. Forest Pathology: Unique features and prospects. *Annu. Rev. Phytopathol.* 12, 431-450.

Sinclair WA, Hudler GW. 1988. Tree declines: Four concepts of causality. J Arboric 14(2), 29–35.

Sinclair WA, Lyon HH. 2005. Diseases of Trees and Shrubs. Ithaca, New York: Cornell University Press. 660 pp. 2nd ed

4 | Forest Tree Pathogens

Forest etiology, which deals with the causes of diseases, either infectious agents (biotic), such as bacteria, viruses, and fungi, or noninfectious agents (abiotic), such as environmental factors including heat, cold, nutrient deficiencies, air pollution, and genetically induced factors. Like all plants, forest trees are subject to injury and disease caused by adverse environmental influences, including other organisms. They may be affected at all stages in their life cycle, from seed to mature tree. Diseases produce a variety of effects and can cause losses in economic, environmental, recreational, and aesthetic values produced by the forest. In forestry, tree diseases are caused by a diverse range of organisms, including fungi, oomycetes, bacteria, phytoplasmas, parasitic higher plants, viruses and nematodes. Among these, the fungi are the largest group causing disease in forest trees. Different *Phytophthora* spp. belonging to the oomycetes group is also important as the causal agent of several devastating diseases on trees especially under nursery.

A pathogen is an agent that causes disease ("path"-"gen", literally disease-generator). It can be living (fungus) or non-living (pollutant). In common parlance, it is often used just to mean living agents. The term doesn't really apply to deficiencies. Here are the main kinds of biotic pathogens:

1. Oomycetes
2. Fungi
3. Bacteria
4. Phytoplasma

5. Virus
6. Algae
7. Plant parasitic nematodes
8. Parasitic flowering plants

General Characters of Tree Pathogens

1. Oomycete

The oomycetes are known as "water/ slime molds", are a group of numerous hundred organisms that comprise some of the most distressing plant pathogens. They are fungus-like eukaryotic microorganisms, filamentous, heterotrophic, and can reproduce both sexually and asexually. Because of their filamentous habit, parasitic oomycetes (kingdom Stramenopila) traditionally have been considered together with fungal pathogens, although they are more closely related to chrysophyte algae. Relatively few oomycetes are associated with foliage diseases of trees; those that cause tree diseases mainly attack roots. The diseases they cause include seedling blights, damping-off, root rots and foliar blights.
Examples: *Albugo, Aphanomyces, Pythium, Phytophthora* etc.

2. Fungi

Fungi are eukaryotic, spore-bearing, achlorophyllous organisms that generally reproduce sexually and asexually and whose filamentous, branched somatic structures are typically surrounded by cell walls consisting of chitin or cellulose or both with many organic molecules. This is the largest group of forest pathogen. The filamentous fungi have advanced a range of structural variations to breach directly and colonize plant tissues and cells to form parasitic relationships. Fungal pathogens can also invade host tissue through wounds caused by breakage, insect feeding and freezing injury.

Symptoms produced by foliar fungal pathogens can vary in harshness from severe defoliation to inconspicuous leaf spots. Many have prolonged incubation periods during which infected foliage remains symptomless. Symptoms of fungal foliage diseases include leaf spots, leaf blotch, scorch, anthracnose, needle blights, and needle casts. Typically, the presence of a fungal pathogen is indicated by minute fruiting structures characteristic of the pathogen species or taxonomic group. Fungal foliage diseases rarely cause the death of mature trees, although they can be responsible for serious economic losses due to decreased growth yield. Chronic fungal foliage diseases can weaken or stress trees so that affected hosts are outcompeted by other species, or are predisposed to attack by secondary agents such as bark beetles or opportunistic root diseases. Management of fungal diseases in forest plantations by protectant fungicides is feasible for certain pathogens and hosts. Successful control depends on the timing of applications to coincide with optimum periods of pathogen sporulation and host susceptibility.

The highest diversity of foliar and wood rot fungal pathogens of tree hosts is ascomycetes (phylum Ascomycota) and basidiomycetous (order Uredinales). Symptoms of fungal diseases include anthracnose, needle blights, needle casts, leaf blotch, leaf spots, scorch, and rust. Naturally, the occurrence of a fungal pathogen is visualized by minute fruiting structures distinctive of the pathogen species or taxonomic group.
Examples: *Armillaria*, *Fusarium*, *Ganoderma*, *Erysiphe*, *Melampsora*, *Rhizoctonia* etc.

3. Bacteria

Bacteria are prokaryotic microscopic organism of the kingdom Eubacteria, rigid, extremely minute, fundamentally unicellular organisms free of green chlorophyll pigments, most commonly multiplying asexually by simple transverse fission, the resulting cell, being of equal or nearly equal in size and budding of daughter cells.

Bacteria are unable to penetrate plants on their own; bacteria require natural openings such as lenticels, stomata, hydathodes, and leaf scars; or wounds due to events such as pruning, breakage, herbivory, or insects. Local dispersal of bacteria from an infected tree occurs via water splash, wind, insects, and rubbing together of infected and healthy foliage. Long-distance dispersal is primarily through the transport of infected material such as seed, stools, and cuttings. Although bacterial pathogens are widespread, information on bacterial diseases of forest trees is limited. In general, bacterial diseases of trees in combined forest systems are less, often because they occur in low incidence or in more severe but sporadic outbreaks, during which they can cause heavy losses locally. The diseases they cause include a shot hole, foliar blight and wilts.
Examples: *Erwinia*, *Xanthomonas*, *Ralstonia* etc.

4. Phytoplasma

Phytoplasmas are obligate parasites, structurally pleomorphic bacteria of the class Mollicutes, wall less prokaryotic plant pathogenic microorganism that can infect living plants. Phytoplasmas are not been grown in artificial culture. Because of this, they have not been fully characterized and their pathogenic nature has not classically been confirmed, although it is assumed. Until the plant-inhabiting pleomorphic mollicutes can be cultivated and characterized apart from their hosts, they are provisionally classified as Candidatus Phytoplasma, with species delineated by genome size and phylogeny as inferred by the nucleic acid sequence of the 16S ribosomal RNA.

Symptoms with which phytoplasmas have been associated may include several of the following: spike, leaf stunting (little leaf), yellow discoloration of the foliage (yellows), greening of flower petals, proliferation of shoots or branches (witches' brooming); loss of apical dominance of the shoots, phloem necrosis, branch dieback and eventual plant death.
Examples: *Candidatus* Phytoplasma pini

5. Virus

Viruses are sub-microscopic, obligate mesobiotic parasite consisting of nucleic acid and protein that multiplies only intracellularly in living cells. Viruses are transmitted mechanically, by contact of infected plant material with healthy tissues, by insect, fungal, or nematode vectors, pollen and seed. Plant viruses cannot be directly controlled by chemical applications.
Examples: *Bamboo mosaic potexvirus*, *Oak mosaic virus*, *Elm mottle virus*, *Prunus necrotic ringspot virus* etc.

6. Algae

Algae are eukaryotic, uni or multicellular organisms, aerophilic, photosynthetic, filamentous green algae containing chlorophyll and orange carotenoid pigments cause plant diseases. They require a thin film of water to complete their life cycles. The genus *Cephaleuros* is a prime member of the Trentepohliales and a unique order, Chlorophyta, which contains the photosynthetic organisms known as green algae. Regular rains and warm weather are favorable conditions for this pathogen. Host infection begins with poor soil drainage, poor plant nutrition, and stagnant air, which are influencing factors to infection by the algae. Symptoms and crop damage can vary greatly depending on the combination of *Cephaleuros* species, hosts and environments.
Examples: *Cephaleuros*

7. Plant parasitic nematodes

Nematodes are thread-like roundworms that live in a wide range of environments including soil and fresh and salt water. Nematodes are harmful plant pathogens causing serious yield loss in a wide range of plants and crops. Nematodes nourish on all parts of the plant, including

leaves, stems, roots, flowers and seeds. Nematodes feed from plants in a variety of ways, but all use a specific spear called a stylet. Their attacks on fine feeder roots of trees can lead to browning, deformity, and death of roots or the production of root galls. Direct damage to plant tissues by shoot-feeding nematodes includes reduced vigor, distortion of plant parts, and death of infected tissues.
Examples: *Meloidogyne*, *Xiphinema*, *Macroposthonia*, *Helicotylenchus*, *Tylenchorhynchus Pratylenchus*, *Trichodorus* and *Paratrichodorus* etc.

8. Parasitic flowering plants

Parasitic flowering plants as an angiosperm (flowering plant/ dicotyledonous seed plants), that directly attaches to the host plant via a haustorium. A haustorium is a specialized attachment and anchoring structure that forms a morphological and physiological link between the parasitic plant and host tree. Host tree responses to colonization include the proliferation of abnormal branches in clusters called "brooms," dieback of branches, and eventual tree death.
Examples: *Arceuthobium* sp., *Dendrophthoe* sp., *Phoradendron* sp., *Viscum* sp.

Major Indigenous and Invasive Tree Pathogens in the Current Era

1. *Armillaria mellea* is a common worldwide pathogen of trees, woody shrubs and herbaceous plants causing root rot, root-collar rot and butt rot. A natural component of forest ecosystems, it can cause wood decay, growth reduction and even mortality, particularly in trees stressed by other factors, or in young trees planted on sites from which infected hosts have been removed.

2. *Botryosphaeria dothidea* is known in many parts of the world and is commonly associated with cankers and dieback on hundreds of different woody plants, including eucalypts. Eucalypt species are planted around the world which makes this species a major concern to the forest sector. This pathogen is an opportunistic pathogen that attacks stressed trees. The entire taxonomy of the genus Botryosphaeria has undergone changes in the last few years. This species was previously considered synonymous with Botryosphaeria ribis, so literature on both species is inextricably intertwined, but these are now considered two distinct species.

3. *Chrysoporthe cubensis* previously named *Cryphonectria cubensis*, is a widespread fungus well known for the canker disease it causes, particularly in *Eucalyptus* spp. The cankers result in limb and trunk breakages, stunted and distorted growth, and often mortality. This is a very important disease in eucalyptus plantations as it is known to kill significant

numbers of trees, particularly those in young plantations, and as a result it is a major constraint to the successful establishment of eucalypt plantations.

4. *Mycosphaerella pini*, 1957 is a fungus that infects and kills the needles of *Pinus* spp. resulting in significant defoliation, stunted growth and eventually death of host trees. It is a major pest in both naturally regenerated and planted forests and probably the most important foliage disease of exotic pines. Susceptibility among pine species does vary. The widely planted *P. radiata* is particularly susceptible and many forests planted with this species in the Southern Hemisphere, particularly in East Africa, New Zealand and Chile, have been devastated by this needle blight. This pathogen has forced managers in some areas to abandon the planting of *P. radiata* and depend more on other tree species.

5. *Phytophthora ramorum* causes a very serious disease called sudden oak death which causes extensive mortality of tanoak and oaks. It is also associated with disease on ornamental plants and other broadleaf and conifer trees. This pathogen has been a significant problem in North American and European forests and nurseries.

List of major tree pathogens in the current era

S. No	Pathogen	Phylum; Order; Family	Country/ Region	Host type
1.	*Armillaria mellea*	Basidiomycota: Marasmiaceae	Kenya, Malawi, Sudan, China, India, Kyrgyzstan	Broadleaf, Conifer
2.	*Botryosphaeria dothidea*	Ascomycota: Botryosphaeriaceae	South Africa, Thailand, Uruguay	Broadleaf
3.	*Chrysoporthe cubensis*	Ascomycota: Incertae sedis	Thailand, Brazil, Colombia	Broadleaf
4.	*Coniothyrium zuluense*	Ascomycota: Leptosphaeriaceae	Thailand, Uruguay	Broadleaf
5.	*Nattrassia mangiferae*	Ascomycota: Incertae sedis	Thailand, Sudan	Broadleaf
6.	*Phanerochaete salmonicolor*	Basidiomycota: Phanerochaetaceae	South Africa, Brazil	Broadleaf
7.	*Mycosphaerella pini*	Ascomycota: Mycosphaerellaceae	Kenya, South Africa	Conifer

8.	*Sphaeropsis sapinea*	Ascomycota: Incertae sedis	Kenya, South Africa	Conifer
9.	*Agrobacterium tumefaciens*	Rhizobiales: Rhizobiaceae	China, Indonesia	Broadleaf
10.	*Microsphaera abbreviata*	Ascomycota: Erysiphaceae	Moldova, Romania	Broadleaf
11.	*Ceratocystis fimbriata*	Ascomycota: Ceratocystidaceae	Brazil, Uruguay	Broadleaf
12.	*Puccinia psidii*	Basidiomycota: Pucciniaceae	Brazil, Uruguay	Broadleaf
13.	*Chrysoporthe cubensis*	Ascomycota: Incertae sedis	Brazil, Colombia	Broadleaf
14.	*Ophiostoma ulmi*	Ascomycota: Incertae sedis	North America, Canada	Conifer
15.	*Cronartium ribicola*	Ascomycota: Pucciniaceae	North America, Canada	Broadleaf
16.	*Phytophthora ramorum*	Oomycota: Peronosporaceae	North America, Canada	Broadleaf
17.	*Nectria coccinea* var. *faginata*	Ascomycota: Nectriaceae	North America, Canada	Broadleaf
18.	*Sirococcus clavigignenti-juglandacearum*	Ascomycota: Incertae sedis	North America, Canada	Broadleaf
19.	*Subramanianospora vesiculosa*	Ascomycota: Incertae sedis	Mauritius, India, Thailand	Broadleaf
20.	*Balansia linearis*	Ascomycota: Clavicipitaceae	India	Broadleaf

Four of the pathogenic species were reported as both indigenous and introduced as the fungal pathogens *Armillaria mellea, Botryosphaeria dothidea, Nattrassia mangiferae* and *Subramanianospora vesiculosa* in India

Selected References

Andrivon D. 1995. Nomenclature for pathogenicity and virulence: Precision vs. tradition. Phytopathology 85, 518-519.

Büttner C, von Bargen S, Bandte M, Mühlbach HP. 2013. Forest diseases caused by viruses. In: Infectious forest diseases. Gonthier P., Nicolotti G. (eds), CABI, p. 50-75.

Early R, Bradley BA, Dukes JS, Lawler JJ, Olden JD, Blumenthal DM, Tatem AJ. 2016. Global threats from invasive alien species in the twenty-first century and national response capacities. Nature Communications, 7, 12485.

FAO (Food and Agriculture Organization of the United Nations). 2009. Global review of forest pests and diseases. FAO Forestry Paper, 156, 222 pp, ISBN 978-92-5-106208-1.

Hepting GH, Cowling EB. 1977. Forest Pathology: Unique features and prospects. *Annu. Rev. Phytopathol.* 12, 431-450.

Schroth G, Krauss U, Gasparotto L, Aguilar JD, Vohland K. 2000. Pests and diseases in agroforestry systems of the humid tropics. Agroforestry Systems, 50(3), 199-241.

Sinclair WA, Lyon HH. 2005. Diseases of Trees and Shrubs. Ithaca, New York: Cornell University Press. 660 pp. 2nd ed

Wingfield MJ, Swart WJ. 1994. Integrated management of forest tree diseases in South Africa. Forest Ecology and Management, 65(1), 11-16.

5 Forest Tree Disease Epidemiology

Forests are the prevailing and most dynamic terrestrial ecosystems on our earth, and despite all of the cumulative knowledge on the forest diseases that threaten them, every year a number of novel 'pathosystems' are revealed as a result of the introduction of exotic pathogens, hybridization, host shifts and jumps, and recombination among pathogens, etc. Additionally, current predictions on trends caused by climate change suggest that the incidence of some diseases will rise, while some may decline (Pautasso et al., 2012); but as the climate becomes less optimal for forests, it is inevitable that we will witness an emergence of diseases that currently are regarded as insignificant or secondary.

Epidemiology refers to the study of the occurrence and cause of infectious diseases, their origin, and their distribution in space and time within a population (Schumann, 1991). It is referred to diseases among a human population, but the term is also widely used for the spread of diseases in plant populations. The term epidemic is habitually misused to describe only rapid and widespread infections, but it should refer to the progress of any disease, irrespective of its speed and extent (Tainter and Baker, 1996). Forest disease epidemiology deals with the incidence, distribution, and development of infectious tree diseases in populations, including populations of infectious pathogens, populations of susceptible hosts, and the interactions of these populations under varying favourable environmental conditions.

Here are the types of effects that come under the preface of epidemiology:

- Mathematical analysis of disease over time in populations.
- Study of factors affecting rate of disease progress: time of leaf wetness, weather, plant development, etc.

» Modelling of disease epidemics. It allows one to vary starting conditions and see the effects.

» Prediction of the effects of disease management methods.

» Monitoring for infection conditions to determine when fungicides or other measures are necessary.

Tree Epidemics

The periodic outbreak or epidemic, and even the severe, sudden, and long-lasting damage resulting from introduction of exotic pathogen, are exceptions rather than the rule in forests. The vast majority of beneficial microorganisms and pathogens have long evolved with the tree species comprising the forests they inhabit, and only a relatively small proportion of these are capable of causing severe damage. There are plentiful examples of forest tree disease epidemics that have caused devastating harms to natural ecosystems. Most of these have happened on trees in natural ecosystems into which pathogens have been introduced. Generally, these pathogens occur on the same genera of trees in areas, separated geographically from those that are affected. Typical examples are Dutch elm disease caused by *Ophiostoma ulmi* and *O. novo-ulmi*, chestnut blight caused by *Cryphonectria parasitica*, white pine blister by *Cronartium ribicola*, and pine wilt caused by the pine wood nematode (*Bursaphelenchus xylophilus*). The first recorded forest disease epidemics date back to the beginning of the 12th century. That was also the time when worldwide trade in and movement of forest products began to upsurge. Today there are accumulative numbers of tree disease epidemics appearing. This is likely to be a trend that will continue, with the increasing movement of people and products around the sphere.

Major outbreak of forest diseases around world

S. No.	Disease	Pathogen	Year and Country
1.	White pine blister rust	*Cronartium ribicola*	1906 in North America
2.	*Annosum* root disease	*Heterobasidion annosum*	1909 in western United States
3.	Dutch elm disease	*Ophiostoma ulmi*	1910 in Netherland
4.	*Sphaeropsis* shoot blight	*Sphaeropsis sapinea*	1917 in New Jersey
5.	Elm yellows	*Candidatus* Phytoplasma ulmi	1919 in Ohio
6.	Swiss needle cast	*Phaeocryptopus gaeumannii*	1925 in Switzerland

7.	Larch canker	*Lachnellula willkommii*	1927 in Massachusetts
8.	Birch dieback	*Melanconium betulinum*	1930 in United States
9.	Oak decline	*Phytophthora cinnamomi*	1930 in Mexico
10.	Littleleaf	*Phytophthora cinnamomi*	1934 in Alabama
11.	Fusiform rust	*Cronartiium quercuum* f. sp. *fusiforme*	1940 in United States
12.	Pitch canker	*Fusarium circinatum*	1946 in Mexico
13.	Charcoal canker	*Biscogniauxia mediterranea*	1960 in Italy
14.	Dutch elm disease	*Ophiostoma novo-ulmi*	1960 in Eastern Europe and North America
15.	*Dothistroma* needle blight	*Dothistroma septosporum* and *Dothistroma pini*	1980s in Central America
16.	Poplar leaf spot	*Marssonina brunnea*	1981 in western United States
17.	*Septoria* canker	*Septoria musiva*	1990 in North America
18.	Ash dieback	*Hymenoscyphus fraxineus*	1990 in East Asia
19.	Pieris blight	*Phytophthora ramorum*	2001 in Germany
20.	Olive quick decline syndrome	*Xylella fastidiosa*	2013 in Italy

Disease Triangle

The disease triangle is a scale-free, flexible, and valuable experimental tool for predicting and analyzing drivers of disease emergence that has been used extensively to determine the etiology of many plant diseases. The prototypical is a conception of host, pathogen, and environment in a tripartite active interaction outline that has been especially useful when realistic at the individual-to-stand level. However, this solicitation is in some contrast to the scale of the most problematic disease outbreaks which tend to occur across landscapes or regions. Forest disease epidemics progress as the eventual consequence of the interaction between a susceptible host plant, a virulent pathogen and a favourable climate. The interaction between these three factors of plant diseases, i.e. the host plant, the infectious agent and the environment, can be envisaged and defined by the disease triangle.

Factors of Disease Epidemics

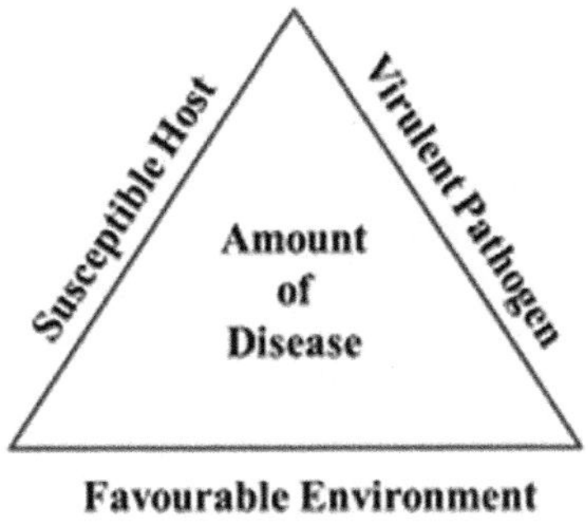

Pathogen

The foremost disease-causing pathogens of forest trees signify many branches on the tree of life, including oomycetes, fungi, bacteria, phytoplasma, virus, nematode or parasitic plants. These pathogens can vary extensively in their spatial distribution across a landscape. Some weak or facultative pathogens can be so extensive as to be nearly omnipresent, likely an evolutionary concern of minimal impacts to host health or primarily saprotrophic energy acquisition. The following pathogen factors, influences the epidemic disease development in forest vegetation.

» Virulence of the pathogen

» Inoculum type

» Inoculum density

» Dispersal potential of the pathogen

» Inoculum potential of the pathogen

» Alternation between hosts

» Adaptation towards host and environment

» Vector compatibility of the pathogen

Host

Disease occurrence or dominance may be obsessed by shifts in host presence or abundance. In forests, species richness reflects a series of climate drivers, ecological progressions, and cultural practices; several examples of documented or possible increases in disease with changes in host population are worth emphasizing in light of these dynamics. Hosts also experience changes in susceptibility due to ecological responses at cellular-to-individual scales. Environmentally driven rises in host susceptibility to pathogens are commonly termed host "predisposition" to infection or disease. Tree hosts respond to physical or biotic stresses such as fire, drought, or insect and pathogen attack concluded a range of chemical defenses, physical modifications, and biochemical pathways. The physiological status of individual trees or whole populations can also determine the likelihood of disease emergence by the following factors

» Resistance of trees

» Susceptibility of trees

» Age of trees

» Uniformity and arrangement of trees

» Population density of trees

» Natural openings

» Wounds and damages

Environment

Environmental factors act on pathogens and tree hosts individualistically and, as a consequence, disease incidence may increase or decrease due to environmental variability. This environmental variation can include annual climate variation (temperature, precipitation, humidity), environmental pollutants (ozone, acid deposition), and a range of biophysical or biotic disturbances such as wind, fire, herbivory or insect outburst. Within the context of the disease triangle, many environmental factors influence disease by limiting hosts or pathogen ranges and through direct influences on pathogens or their hosts are follows

» Temperature

» Humidity

» Rainfall

» Frost

» Drought

» Leaf wetness period

» Fire damage

» Soil fertility

» Soil moisture content

» Soil temperature

Selected References

» Boyce JS. 1961. Forest Pathology, 3rd edn. McGraw-Hill, New York.

» Capretti P, Ragazzi A. 2009. Elementi di Patologia Forestale. Pàtron Editore, Bologna, Italy.

» Edmonds RL, Agee JK, Gara RI. 2011. Forest Health and Protection, 2nd edn. Waveland Press, Long Grove, Illinois.

» Pautasso M, Döring TF, Garbelotto M, Pellis L, Jeger MJ. 2012. Impacts of climate change on plant diseases – opinions and trends. European Journal of Plant Pathology 133, 295–313.

» Sturrock RN, Frankel SJ, Brown AV, Hennon PE, Kliejunas JT, Lewis KJ, Worrall JJ. Woods AJ. 2011. Climate change and forest diseases. Review. Plant Pathology 60, 133–149.

» Tainter FH, Baker FA. 1996. Principles of Forest Pathology. John Wiley, Hoboken, New Jersey.

6 Principles of Forest Tree Disease Management

Management of tree diseases in natural forests and plantations is extremely difficult and in some cases practically impossible. This is very unlike to disease management in plantations where many choices exist for disease prevention. The major practicable methods for forest-disease management are silvicultural practices aimed at avoiding disease losses and using genetically disease-resistant selections of trees. Management recommendations for minimizing disease losses in forests must be

a. Highly flexible
b. Compatible with conventional management goals
c. Packaged for an extension clientele that ranges from small farmers and truant land owners to sensible managers of huge industrial enterprises.

Disease management decisions are not simple determinations. Three components are involved in the synthesis and application of these efforts:

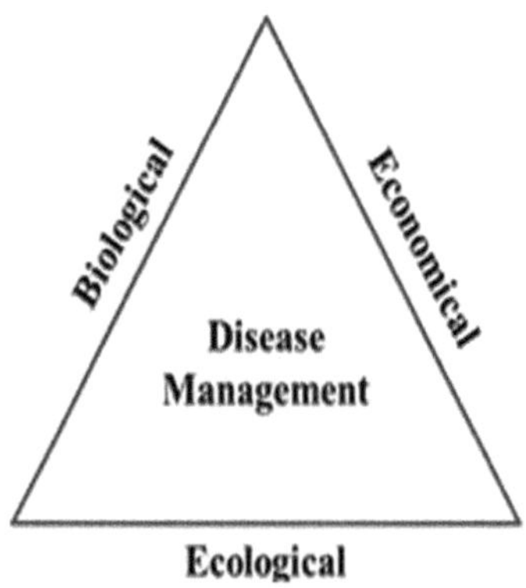

General concepts of plant disease control including the avoidance or exclusion, eradication, protection, and immunization, so widely and effectively used with agricultural crops, are applicable in forest practice. The first two are focused against the causal agent or disease incitant; the last two toward the host plant.

a. **Direct control measures**, specifically framed to limit or even to eradicate a specific pathogen, included temporary or emergency handlings directed against newly introduced pathogens before they become decisively established or against prevailing diseases that have reached, or threaten to reach, epidemic proportions under particularly favorable circumstances. Recent researches have suggested possible uses of antagonistic micro-organisms, antibiotics, growth regulators, systemic chemotherapeutants, and silvicides in tree disease control.

b. **Indirect control measures** attempt to correct or to avoid fundamental disturbances in forest conditions from which disease outbreaks receive their initial stimulus. Such events rely on disease avoidance rather than cure and are applied over a long period of time. Their immediate costs must be balanced against the losses which they prevent and the returns they ensure.

General Concepts of Disease Control

A. Exclusion

One of the important mechanisms of tree disease management lies in excluding pathogens from new or introduced environments. As mentioned previously, some of the most devastating diseases have been caused by pathogens introduced into new environments. The exclusion of pathogens from new environments depends on quarantine regulations that ensure that they are not moved across borders. Contained in the exclusion category are all control measures that function to prevent the entrance and establishment of biotic disease agents into a particular area. These procedures are valid only when the area to which they are directed is free of the pathogen, and they are most relevant in preventing the intercontinental spread of pathogens.

Procedures include the following:

1. **Selection and treatment of propagative materials**: Selecting or treating propagative materials is very difficult in the absence of symptoms and signs. However, seeds may be surface sterilized with hot water or gases.

2. **Quarantines**: The quarantine method is predicated on two primary observations:

a. Comparatively few biotic disease agents are cosmopolitan in distribution, and

b. In the absence of continuous land surfaces and continuous plant distributions, the spread of many pathogens is denied or restricted. However, natural quarantine blocks have been repeatedly broken by humans in a world in which intercontinental travel is commonplace.

3. **Investigation of exotic disease agents**: To exclude exotic pathogens by investigative procedures. Two investigative procedures are

   a. To grow native species in foreign lands and

   a. To import disease agents for intensive study. Also, strict regulations are required to prevent escape of either hosts or pathogens.

B. Eradication

The second strategy used to minimize the impact of the disease is through eradication of pathogens after they have been introduced into new areas. This approach is complicated and depends on early recognition of the disease, a clear understanding of the biology of the causal agent, and intensive efforts to eliminate it. In exclusion, the primary objective is to ensure that a disease agent is not introduced into an uncontaminated area. In eradication efforts, however, the host is infected or the pathogen is present, and the basic purpose of the control procedure is to rid the host or the area of the disease agent.

Four procedures are employed:

1. **Destruction of diseased individuals**: Destroying diseased specimens has been widely applied to combat tree maladies, particularly those in which disease agents move slowly.

2. **Removal of the alternate host**: Many rust fungi require alternate hosts to complete their life cycles. Removal of the least desired host will prevent completion of the life cycle of the pathogen and control the disease in the desirable host. An extension of this procedure would be to remove the "alternative" host if it is of low value. Several fungal pathogens are maintained on weeds, even though the weeds are not necessary for the completion of the life cycle of the fungus.

3. **Therapy**: Therapy may include physical therapy such as surgery, environmental modification, radiation, or chemotherapy such as the application of topical and systemic chemicals.

4. **Disinfestation of the environment**: Disinfestation is usually restricted to greenhouse

and nursery situations, and involves sterilization of the soil. One problem encountered with the procedure, particularly with trees, is the inadvertent destruction of necessary mycorrhizae.

C. Protection

Protective control procedures are those that are employed when a stress factor cannot be excluded or eradicated from an area containing vulnerable plants. Protection is employed only when the plants are uninfluenced (i.e., free of abiotic stress factors) or uninfected (i.e., free of biotic stress agents), but where the possibility of undesirable consequences exists. Four primary control practices of providing protection include the following:

1. **Environmental modification**: Fans and portable heaters can be used to combat cold temperatures. This procedure is not applicable to forest stands, but it is useful in greenhouse or nursery settings and has been effective in citrus plantations.

2. **Silvicultural practices**: Site selection is important with the knowledge that certain species are best adapted to specific soil, slope, and compass direction conditions. Also, selective removal is advocated and effective for certain diseases.

3. **Biological protection**: Biological control of plant diseases is the suppression of populations of plant pathogens by living organisms. Also, of serious concern is the possibility that the biological control agent for a particular pathogen may end up becoming a pathogen itself, either on the host it is presumably protecting or some other species in the vicinity.

4. **Chemical protection**: Chemical control or protection to reduce the impact of tree diseases is not commonly used in forestry. This is due to the fact that forests represent extensive, ecologically sensitive environments where most fungicides cannot be used responsibly. Chemical protection has been one of the most commonly used control strategies. A major concern, of course, is that some pesticides kill nontarget organisms, and some persist in the environment.

D. Resistance

Most living organisms are resistant to maximum pathogens. For example, elm trees do not pact hog cholera, and humans do not agonise from Dutch elm disease. Similarly, oaks do not sussceptible to white pine blister rust, and aspens do not get infected by chestnut blight. The reasons are simple. Organisms have advanced structures and specialized mechanisms over thousands of years that provide resistance to invading pathogens. However, many pathogenic agents are opportunists, and although barriers such as bark and leaf cuticle may offer resistance to these agents, the pathogens may become infective if barriers are broken by injury. Even then, many plants have internal mechanisms for warding off disease, once a potential pathogen has entered. Resistance is perhaps the most environmentally acceptable control measure because it incorporates a host's own genetic code in offering resistance. No chemicals need to be applied, no environmental modification is necessary, and no labor is required. What is required is a procedure for propagating resistant hosts.

Three techniques are used:

1. **Selection**: The selection technique involves the search for individuals of a particular species that appear to be naturally resistant to particular diseases, while other individuals of the same species are not. Propagation of these "apparently resistant" individuals is then pursued, with the expectation that natural selection has produced resistant characteristics that can be used on to future generations.

2. **Breeding**: The goal of agriculture is the constant production of the maximum yields and the utmost quality of the most desirable crop at the lowest cost. The techniques in plant improvement have played a major role in attaining this goal. By genetic modifications of specific characteristics, plant breeders have incorporated favourable genes into thousands of plant varieties. Included in these beneficial genes are genes for resistance that may produce physical or chemical barriers to pathogen infection. Contemporary techniques in molecular biology have improved the guesswork in the breeding process: specific genes can now be inserted directly into the DNA of the host.

3. **Avoiding monocultures and other narrow-genetic lines of susceptible plants**: Most epidemics of tree diseases could be avoided if only a small population of a single species were planted instead of a monoculture. This may not be economically feasible in a nursery or commercial forest plantation, but it should be considered in forest recreation areas and in all urban sites. In addition, the number of trees within a plant family also should be limited. Current guidelines suggest that a tree population contain no more than 5% of anyone species and no more than 10% of species within a single plant family, whenever possible.

Selected References

Bingham RT, Hoff RJ, McDonald GI. 1971. Disease resistance in forest trees. Annu. Rev. Phytopathol. 9, 433-452.

Blanchard RO, Tattar TA. 1981. Field and Laboratory Guide to Tree Pathology, Academic Press, pp 302.

Cook RJ, Baker KF. 1983. "The Nature and Practice of Biological Control of Plant Pathogens." APS Press, St. Paul, Minnesota.

Delp CJ. 1988. Fungicide Resistance in North America. APS Press, St. Paul, Minnesota.

Gonthier P, Nicolotti G. 2013. Infectious forest diseases. CABI, pp 592

Hickey KD. 1986. Methods for Evaluating Pesticides for Control of Plant Pathogens. APS Press, St. Paul, Minnesota.

Leonard KJ, Fry WE. 1986. Plant Disease Epidemiology. Macmillan, New York.

Liebhold AM, MacDonald WL, Bergdahl D, Mastro VC. 1995. Invasion by exotic pests: A threat to forest ecosystems. For. Set Monogr. 30, 1-49.

Phillips DH, Burdekin DA. 1982. Diseases of Forest and Ornamental Trees. Palgrave Macmillan UK, pp 463.

Smith WH. 1970. Tree Pathology: A Short Introduction. Academic Press, pp 324.

Tattar TA. 1989. Diseases of Shade Trees, Revised Edition. 1st Edition, Academic Press, pp 391.

Van Der Plank JE. 1964. Plant Diseases: Epidemics and Control, Academic Press, New York.

Wolf MS. 1985. The current status and prospects of multiline cultivars and variety mixtures for disease resistance. Annu. Rev. Phytopathol. 23, 251-273.

7 Diseases of *Acacia* sp. (Acacia)

(*Acacia aulacocarpa, A. crassicarpa, A. mangium* and *A. auriculiformis*)

S. No.	Disease	Pathogen
1.	*Alternaria* tip necrosis	*Alternaria* sp.
2.	*Pestalotiopsis* tip necrosis	*Pestalotiopsis* sp.
3.	*Colletotrichum* tip necrosis	*Colletotrichum* sp.
4.	*Phomopsis* tip necrosis	*Phomopsis* sp.
5.	*Curvularia* tip necrosis	*Curvularia* sp.
6.	Unidentified tip necrosis	Hyphomycete
7.	*Colletotrichum* foliar spot	*Colletotrichum acacia*, *Colletotrichum gloeosporioides*
8.	*Alternaria* foliar spot	*Alternaria alternata*
9.	*Phomopsis* foliar spot	*Phomopsis* sp.
10.	*Pestalotiopsis* foliar spot	*Pestalotiopsis* sp.
11.	*Glomerella* foliar spot	*Glomerella cingulata*
12.	*Cylindrocladium* foliar spot	*Cylindrocladium quinqueseptatum*, *Cylindrocladium scoparium*
13.	*Sporidesmium* foliar spot	*Sporidesmium australiense*
14.	*Phaeotrichoconis* foliar spot	*Phaeotrichoconis* sp.

15.	*Mycosphaerella* foliar spot	*Mycosphaerella acaciigena, Mycosphaerella thailandica* and *Pseudocercospora* sp
16.	*Curvularia* foliar spot	*Curvularia* sp.
17.	*Cercosporidium* foliar spot	*Cercosporidium acacia*
18.	*Phoma* foliar spot	*Phoma* sp.
19.	*Macrophomina* foliar spot	*Macrophomina* sp.
20.	*Exoserohilum* foliar spot	*Exoserohilum rostratum, Phaeotrichoconis crotalariae*
21.	Unidentified foliar spot	Hyphomycete
22.	Die back	*Hendersonula* sp.
23.	Root rot	*Ganoderma philippii, Ganoderma mastoporum, Ganoderma aff. steyaertanum, Ganoderma australe, Ganoderma mastoporum, Ganoderma cupreum, Ganoderma sinense* and *Amauroderma rugosum*
24.	Heart rot	*Phellinus pachyphloeus, Trametes palustris, Tinctoporellus epimiltinus, Rigidoporus hypobrunneus, Schizophyllum commune, Fomes noxius, Polyporus alveolaris*
25.	Powdery mildew	*Oidium* sp., *Erysiphe acaciae, Erysiphe quercicola*
26.	Black mildew	*Meliola brisbanensis*
27.	Pink disease	*Phanerochaete salmonicolor*
28.	Seedling blight	*Colletotrichum gloeosporioides*
29.	Damping-off	*Rhizoctonia solani, Pythium* sp., *Phytophthora* sp.
30.	Leaf web blight	*Rhizoctonia solani*
31.	Stem canker	*Nectria* sp.
32.	Basal stem canker and root decay	*Botryodiplodia theobromae*
33.	Branch canker	*Cytospora* sp.
34.	Shoot blight	*Phomopsis* sp.
35.	Stem decay	*Gliocladium* sp.
36.	Basal stem decay	Basidiomycete
37.	Root decay and root rot	*Macrophomina phaseolina, Armillaria mellea, Coriolopsis aspera* and *Phellinus* spp. on *A. auriculiformis* and *A. nilotica; Phellinus noxius* on *A. mangium*

38.	Rust	*Uromycladium tepperianum, Uromycladium acacia, Uromyces digitatus, Aecidium* sp.
39.	Pod rot	*Fusicoccum* sp.
40.	Stem malformation	*Ceratocystis manginecans, Ceratocystis acaciivora* on *A. auriculiformis, A. mangium*
41.	Bacterial leaf blight	*Xanthomonas campestris*
42.	Spike disease	*Candidatus* Phytoplasma sp.
43.	Green algal spot	*Cephaleuros virescens*

I. Foliar Tip Necrosis

There were five distinct types of tip necrosis diseases. Since tip necrosis was caused by various fungi producing varied symptoms each disease is described separately.

1. *Alternaria* tip necrosis

Symptoms

Acacia aulizcocarpa

- It is reported in *Acacia aulizcocarpa* in India.
- The disease was characterised by drying up of the phyllode tips.
- Tip necrosis, dark brown in colour, proceeded slowly downwards towards the centre of the phyllode as seen from dark brown zonation lines in the necrotic region.

Acacia crassicarpa

- Initially, small dark tan lesions developed near the tip which gradually extended towards the central portion of the phyllode.
- These lesions coalesced to form large necrotic dark tan coloured areas extending from the tip downwards, more towards the inner margin.
- The affected phyllodes abscissed prematurely.
- Dark brown coloured necrosis started from the tip and extended downwards.
- Characteristic dark grey streaks running parallel to veins, developed in the necrotic area where dark minute fructifications were produced.

Acacia auriculiformis

- The disease affected the younger phyllodes near the growing shoot.
- The infection at the tip caused dark tan necrosis which advanced slowly downwards.
- The growing margin of the necrosis appeared dark brown.

Fungus: *Alternaria* **sp.**

Management

- Application of fungicides such as Carbendazim 0.05% or Chlorothalonil 0.2%.

2. *Pestalotiopsis* tip necrosis

Symptoms

Acacia aulizcocarpa

- The tips of the affected phyllodes were greyish brown on which the fungus sporulated as minute black dots.
- Almost all the phyllodes on the lower branches of most of the trees were affected.
- As the disease was localised at the tip it did not cause any defoliation.

Acacia crassicarpa

- The phyllodes had numerous spots, pale brown in colour with, dark brown centres.
- There were more spots near the tip and middle of the phyllodes than at the base.
- When seen against the light the spots appeared as pale translucent areas with a dark centre.

Fungus: *Pestalotiopsis* **sp.**

Management

- Application of fungicides such as Carbendazim 0.05% or Chlorothalonil 0.1%.

3. *Colletotrichum* tip necrosis

Symptoms

Acacia aulizcocarpa

- The disease was characterised initially by the appearance of a dark brown spot with pale margin near the tip of the phyllode.
- Symptoms gradually spread towards the tip as well as centre of the phyllode.
- The disease mostly affected the mature phyllodes.

Acacia mangium

- Necrosis, dull greyish in colour, was initiated at the tip and progressed downwards, sometimes covering up to one-fourth of phyllodes.
- The necrotic area became brittle and started to crack and break into smaller pieces.
- Occasionally, fructifications of the causal organism were also found over the necrotic tissues.

Acacia auriculiformis

- Initially, the necrosis at the tip was brownish in colour later changing to greyish.
- The infection, localised at the tip of mature phyllodes, appeared as dull greyish necrotic areas with dark brown margins.
- The necrosis advanced further down leaving behind the previous dark brown margin.

Fungus: *Colletotrichum* **sp.**

Management

- Application of fungicides such as Benomyl 0.1% or Tricyclazole 0.1% or Chlorothalonil 0.1%.

4. *Phomopsis* tip necrosis

Symptoms

- The severity of the disease was rated as medium.

- Light greyish necrosis was initiated at the tip which spread gradually down the lamina with irregular raised and well defined dark brown margin.
- Occasionally, the fructifications of the causal organism were also produced over the necrotic area.

Fungus: *Phomopsis* **sp.**

Management

- Application of fungicides such as Carbendazim 0.05% or Chlorothalonil 0.1%.

5. *Curvularia* tip necrosis

Symptoms

- It is reported in *Acacia crassicarpa* in India.
- The infection was mostly localised on lower phyllodes.
- The infection started as greyish light brown spots at the tip, gradually spreading downwards. The rapidly spreading infection caused premature defoliation.

Fungus: *Curvularia* **sp.**

Management

- Application of fungicides such as Carbendazim 0.05% or Chlorothalonil 0.1%.

6.1. Unidentified tip necrosis

Symptoms

- It is reported in *Acacia aulizcocarpa* in India.
- The reddish brown necrotic area initially developed at the tip, gradually spreading downwards, covering almost half of the phyllode.
- The disease was observed on mature phyllodes at were affected with this disease it appeared to be a serious one.

Fungus: Unidentified, non-sporulating

6.2. Unidentified tip necrosis

Symptoms

- It is reported in *Acacia aulizcocarpa* in India.
- The disease appeared to be of a serious nature as the affected phyllodes abscissed prematurely.
- The primary symptom of the disease was the development of water soaked linear lesions along the veins near the tip which soon turned greyish.
- These lesions coalesced to form greyish to light brown large necrotic areas.
- As more and more spots developed, they coalesced to enlarge the necrotic area, characterised by a dark greyish advancing margin or zonation line.
- The affected phyllode became pale and abscissed prematurely.

Fungus: Unidentified, non-sporulating

6.3. Unidentified tip necrosis

Symptoms

- Dull greyish brown necrosis started from the tip and spread down the phyllode in an oblique fashion, covering more than one-fourth of the lamina.
- Numerous greyish black streaks developed in the necrotic area running parallel to the veins on which black fructifications of the causal organism developed.

Fungus: Unidentified, non-sporulating

6.4. Unidentified tip necrosis

Symptoms

- Almost 75% of the phyllode length was covered with a large necrotic area which was characteristically light brown at the tip and dark tan towards the base.
- Within the necrotic area elongated light greyish spots with diffused margins were also seen.

Fungus: Unidentified, non-sporulating

6.5. Unidentified tip necrosis

Symptoms

» Light brown necrotic lesions developed near the tip, which coalesced to form large necrotic spots.

» The spot advanced further affecting the lamina.

» The advancing margin of the spot was defined and lighter in colour.

» Numerous minute brown fructifications developed in the necrotic area.

Fungus: Unidentified; non-sporulating Hyphomycete.

II. Foliar Spots

Seven different types of foliar spot diseases with which different fungal organisms were associated were recorded from various localities.

7. *Colletotrichum* foliar spot

Symptoms

Acacia aulizcocarpa

4. The typical symptoms of the disease were spherical to oval black spots, 1-2 mm diam., scattered on the lamina irrespective of vein arrangements.
5. The tissue around the spots became necrotic and such small necrotic lesions coalesced to form large necrotic areas, which spread from phyllode tips downwards.

Acacia crassicarpa

» The disease was characterised by chocolate brown large necrotic spots, mostly elongated with darker margins.

» The spots were found only on mature phyllodes.

Acacia mangium

» The disease was characterised by initial circular to oval -reddish brown spots of variable sizes with raised defined margins.

- There were more spots in the centre and basal part of the phyllodes.
- Smaller spots coalesced to produce larger irregular spots.
- Severe infection caused premature defoliation.
- The infection was widespread and severe on mature phyllodes.
- The plants were closely spaced providing conducive microclimatic conditions for disease occurrence and spread.

Acacia auriculiformis

- In *A. auriculiformis* black spot was caused by *Colletotrichum gloeosporioides*
- Initially, the spots, dark brown in colour, were 1 mm across.
- These spots developed an elongated light greyish white area and appeared elliptical or spindle shaped.
- There were more spots in the upper surface than the lower.
- Most of the spots were of uniform size.

Fungi: *Colletotrichum acacia, Colletotrichum gloeosporioides*

Management

- Spray with Copper hydroxide 0.25% or Carbendazim 0.05% or Chlorothalonil 0.1% or Mancozeb 0.1%.

8. ***Alternaria*** **foliar spot**

Symptoms

Acacia aulizcocarpa

- The disease was of moderate severity resulting in premature defoliation.
- Initially, dark brown spots appeared more towards the margins of the phyllode.
- Around the spots the tissue became light brown and coalesced giving rise to large and necrotic light brown areas along the margins.
- Severely affected phyllodes turned pale and abscissed prematurely.

Acacia crassicarpa

- » Usually, the necrotic spots started from the outer margin near the tip of the phyllodes and extended downwards towards the base developing into large greyish brown spots, with dark brown margins.
- » Numerous black coloured fructifications developed on both the sides of the spots.

Acacia mangium

- » The spots were light greyish circular to elliptical areas with rough outline measuring 2-3 mm.
- » The spots mostly occurred near the base of the phyllode.

Acacia auriculiformis

- » The disease was characterised by small angular dark purplish spots (1-2 mm across), more on the upper surface of phyllodes.
- » Only when some of the spots coalesced could the spots could be seen on the lower surface.

Fungus: *Alternaria alternata*

Management

- » Spray with Chlorothalonil 0.1% or Mancozeb 0.1%.

9. *Phomopsis* foliar spot

Symptoms

Acacia aulizcocarpa

- » Initially, light discoloured areas developed on any part of the phyllode.
- » At the centre of these areas irregular, small greyish spots with dark brown margin appeared. The greyish spots enlarged in size and covered; up to half the lamina.
- » Severely infected phyllodes abscissed prematurely.

Acacia auriculiformis

- » The amphigenous spots appeared spherical to elliptical dark brown to tan with a pale halo.

- Smaller spots (2-3 mm across) coalesced to form irregular larger spots, measuring 3-4 mm across.

Fungus: *Phomopsis* **sp.**

Management

- Spray with Carbendazim 0.1% or Mancozeb 0.1%.

10. *Pestalotiopsis* foliar spot

Symptoms

Acacia aulacocarpa

- The disease was characterised by irregular, elongated brown coloured spots which spread rapidly to coalesce and form large necrotic areas.
- Severely affected phyllodes defoliated prematurely.

Acacia mangium* and *Acacia crassicarpa

- Initially small dark brown spots coalesced to produce elongated dark brown to tan spots with irregular margins between the veins, occasionally spreading phyllode tips or margins.
- The disease produced typical interveinal dark brown, large elongated spots with rounded/obtuse ends and dark margin only at the apical end.
- The spots enlarged more lengthwise than across; adjacent spots coalesced together to form larger spots.
- The disease mostly occurred on younger phyllodes.
- Numerous black coloured fructifications were produced in the necrotic area.
- Severe infection caused yellowing and premature defoliation.

Acacia auriculiformis

- The disease, mostly occuring on the upper surface of mature phyllodes, was characterised by reddish brown spots, similar to pock marks with depressed centres, on the lower surface the spots appeared as dark brown slightly raised tissue.

» These spots coalesced to give rise to larger necrotic spots, especially near the tip region.

Fungus: *Pestalotiopsis* **sp.**

Management

» Application of fungicides such as Carbendazim 0.1% or Chlorothalonil 0.2%.

11. ***Glomerella*** **foliar spot**

Symptoms

Acacia aulizcocarpa

» The incidence and severity of the disease were low.

» The disease was characterised by irregular greyish brown spots with dark brown margins. Smaller spots coalesced to form large necrotic areas.

» Occasionally, dark coloured fungal fiuctifications of the causal organism were observed in some of the spots.

Acacia mangium

» The disease appeared as circular to oval dark brown spots up to 3 mm in diam.

» The spots were surrounded by a light pale brown halo.

» Occasionally, the spots were marked by veins.

Acacia auriculiformis, Acacia cyanophylla **and** ***Acacia catechu***

» The disease was characterised by amphigenous, dark brownish black, oval to spindle shaped spots with silvery grey to light brownish grey centres.

» The adjacent spots coalesced to form larger spots.

Fungus: *Glomerella cingulata*

Management

» Application of fungicides such as Carbendazim 0.1% or Chlorothalonil 0.2%.

12. *Cylindrocladium* foliar spot

Symptoms

Acacia catechu

- » The leaf spot cause small oval shaped spots that coalesce and result in defoliation in both seedlings sand in young trees.

Acacia auriculiformis

- » The disease was characterised by large reddish brown necrotic areas, usually along the margins and tip. The affected phyllodes abscissed prematurely.

Acacia mangium

- » On young phyllodes, symptoms include greyish water-soaked spots which coalesce and develop into extensive necrotic areas, usually along the margins and tips.
- » At this time large numbers of white spores can be seen on the surface of the phyllodes.
- » Under favourable conditions of high humidity and frequent rainfall, the spots further spread to cover the entire area of the phyllode inducing leaf blight symptoms.
- » Affected phyllodes turn reddish brown and are prematurely shed.

Fungus: *Cylindrocladium quinqueseptatum* (*A. catechu*); *Cylindrocladium scoparium* (*A. auriculiformis*)

It produces cylindrical, long hyaline conidia or spores that are air borne through wind and rain.

Management

- » Selection of disease-resistant plants.
- » Spray with Carbendazim 0.1% or Copper oxy chloride 0.05% or Chlorothalonil 0.2% or Mancozeb 0.1%.

13. *Sporidesmium* foliar spot

Symptoms

- » It is reported in *Acacia aulizcocarpa* in India.

- The disease was characterised by elongated dark brown lesions, 0.5 to 1 mm wide, running parallel to the veins.
- These lesions enlarged and coalesced to produce large dark reddish brown spots, which were usually initiated from the tip of the phyllode.
- Severely affected phyllodes turned pale and abscissed prematurely.

Fungus: *Sporidesmium australiense*

Management

- Application of fungicides such as Carbendazim 0.1% or Chlorothalonil 0.2%.

14. *Phaeotrichoconis* foliar spot

Symptoms

- The disease started by developing circular spot of dark brown colour on leaflets.
- The spots formed necrotic brown areas.
- Conidia have long beak-like appendages.

Fungus: *Phaeotrichoconis* **sp.**

Management

- Application of fungicides such as Carbendazim 0.1% or Chlorothalonil 0.2%.

15. *Mycosphaerella* foliar spot

Symptoms

***Acacia crassicarpa* and *Acacia mangium (Mycosphaerella* sp)**

- The infection occurred on young as well as mature leaves.
- The symptoms were blotches and brown necrotic areas.
- First, dark brown in colour on the tip of phyllode appeared and downwards to the centre of the phyllode seen as dark brown zonation lines in the necrotic region.

Acacia crassicarpa (Pseudocercospora **sp)**

- The symptoms produced by both fungi are similar.
- Phyllodes bear spots, blotches and more extensive reddish-brown necrotic areas with scattered, or more densely arrayed, sporulating structures.

Fungi: *Mycosphaerella acaciigena, Mycosphaerella thailandica* and *Pseudocercospora* sp

Management

- Application of fungicides such as Thiophanate methyl 0.2%, Carbendazim + Mancozeb 0.1%.

16. *Curvularia* foliar spot

Symptoms

- It is first reported in *Acacia mangium.*
- The disease produced irregular shaped reddish brown spots scattered on the phyllode.
- The smaller spots coalesced to give rise to large irregular spots with diffuse margins.
- The infection was also observed at the phyllode tips and margins.

Fungus: *Curvularia* sp.

Management

- Application of fungicides such as Carbendazim 0.1% or Chlorothalonil 0.2%.

17. *Cercosporidium* foliar spot

Symptoms

- It is first reported in *Acacia mangium.*
- The disease was widespread and of serious consequences as it caused premature defoliation.
- Initially, the spots appeared on the lower phyllodes and the infection progressed upwards, causing defoliation of the bottom phyllodes due to extensive spotting.
- Numerous spots appeared as dark reddish brown necrotic areas, circular to oval, 1-2 mm in diam.

- These spots increased in size to 3-4 mm across.
- When the expanding spots coalesced, large size necrotic spots were formed.
- Rarely, the spots also affected the veins.

Fungus: *Cercosporidium acacia*

Management

- Application of fungicides such as Carbendazim 0.1% or Chlorothalonil 0.2%.

18. *Phoma* foliar spot

Symptoms

- It is widely reported in *Acacia flavescens.*
- The disease observed on the upper surface of young as well as moderately mature phyllodes in the middle crown of the tree, was characterised by elongated streak-like white lesions up to 5-8 mm in length.
- Some of the lesions coalesced to give rise to irregular shaped lesions.
- In severe cases the lesions also developed on the lower surface.
- Black dot-like pycnidia were observed in the lesions.

Fungus: *Phoma* sp.

Management

- Application of fungicides such as Tebuconazole 0.1% or Trifloxystrobin 0.2%.

19. *Macrophomina* foliar spot

Symptoms

- Dark brown to purplish brown amphigeneous spots were produced on young as well as mature phyllodes.
- The spots, oval to elliptical to spindle shaped, 2-3 mm in length, were localised mostly at the margin; only a few spots developed on the lamina.

- When the spots joined, long irregular shaped lesions were produced.

Fungus: *Macrophomina* sp.

Management

- Application of fungicides such as Carbendazim + Mancozeb 0.1% or Chlorothalonil 0.2%.

20. *Exoserohilum* foliar spot

Symptoms

Acacia aulizcocarpa

- Characteristic brownish red irregular spots phyllodes which enlarged to form large necrotic lesions of varying size developed near the margins of mature.

Acacia crassicarpa

- Discrete spots, randomly scattered on the phyllode.
- Lesions are oblong or elliptical in shape, reddish in colour with a darkly pigmented margin and paler necrotic tissue toward the centre.

Fungus: *Exoserohilum rostratum, Phaeotrichoconis crotalariae*

Management

- Application of fungicides such as Carbendazim 0.1% or Chlorothalonil 0.2%.

21.1. Unidentified foliar spot

Symptoms

- It is reported in *Acacia aulizcocarpa* in India.
- The disease was characterised by elongated dark coloured streaks parallel to the veins, more towards the outer margin of the phyllodes.
- These streaks enlarged and coalesced to give to raise greyish black lesions which resulted in premature defoliation.

Fungus: Non-sporulating Hyphomycetes.

21.2. Unidentified foliar spot

Symptoms

- It is reported in *Acacia aulizcocarpa* in India.
- The symptoms resembled nutrient deficiency.
- The affected phyllodes had typical elongated faint yellowish streaks parallel to the veins throughout the lamina.
- The spots were clearly visible when the phyllode was observed against light.
- The disease was common on young as well as mature phyllodes

Fungus: Non-sporulating Hyphomycetes.

21.3. Unidentified foliar spot

Symptoms

- It is reported in *Acacia crassicarpa* in India
- Initially reddish brown shining irregular spots appeared on the mature phyllodes.
- As smaller spots developed they coalesced to form larger spots.
- The affected phyllodes turned pale and defoliated prematurely.

Fungus: Unidentified; Sphaeropsidales.

21.4. Unidentified foliar spot

Symptoms

- It is reported in *Acacia crassicarpa* in India.
- Initially reddish brown shining irregular spots appeared on the mature phyllodes.
- As smaller spots developed they coalesced to form larger spots.
- The affected phyllodes turned pale and defoliated prematurely.

Fungus: Unidentified; non-sporulating Hyphomycete.

21.5. Unidentified foliar spot

Symptoms

- » It is reported in *Acacia crassicarpa* in India.
- » The disease was characterised by elongated greyish brown long spots along the outer margin of the phyllodes.
- » The necrotic spot had dark greyish long lesions where the fructifications of the causal organism developed

Fungus: Unidentified; non-sporulating Hyphomycete.

21.6. Unidentified foliar spot

Symptoms

- » It is reported in *Acacia crassicarpa* in India
- » The affected phyllodes had light yellowish orange spots with diffuse margins throughout the laminae.
- » When observed against the light, these spots appeared bright yellowish orange.
- » The disease affected the mature phyllodes but it progressed upwards also affecting the younger ones.
- » The affected phyllodes abscissed prematurely.

Fungus: Unidentified; non-sporulating.

21.7. Unidentified foliar spot

Symptoms

- » It is reported in *Acacia mangium* in India.
- » The disease appeared as minute dark reddish brown to maroon coloured spots, less than 1 mm in diam., spread throughout the lamina.
- » These spots expanded to 1-2 mm diam. and developed pale to light brown colour in the centre; coalesced spots formed larger irregular reddish brown coloured spots.

- Severely affected phyllodes abscised prematurely.
- The disease was very severe on mature phyllodes resulting in premature defoliation.

Fungus: Unidentified; non-sporulating Hyphomycete.

21.8. Unidentified foliar spot

Symptoms

- It is reported in *Acacia mangium* in India.
- Initially, the elliptical spots, about 1 cm long, appeared as light orange blotches with diffuse margins; these blotches when seen against the light appeared as yellowish, water-soaked areas.
- In these blotches small necrotic spots developed in concentric rings which coalesced to form large reddish brown to dark brown necrotic spots with light brown haloes around them; severely affected phyllodes abscissed prematurely.

Fungus: Unidentified; non-sporulating Hyphomycete.

21.9. Unidentified foliar spot

Symptoms

- It is reported in *Acacia mangium* in India.
- Numerous circular dark brown spots, 102 mm diam., appeared in the upper half of the spots expanded and coalesced to form large necrotic areas extending from one margin to the other.

Fungus: Unidentified; non-sporulating Hyphomycete.

22. Dieback

Symptoms

Acacia mangium

- This disease, also known as branch wilt, limb wilt or canker, occurred in a tree where summer heat becomes intense.

- » The incidence of dieback could be combined impacts of site, soil, climatic conditions, insects, and a range of pathogenic fungi.
- » Symptom appeared during summer. Leaves wilted on scattered branches and turned brown, cling to dead twigs or senesced and dropped prematurely.
- » The initial symptoms of dieback were yellowing of phyllodes and subsequent thinning of the upper crown, including the top.
- » Initially, the younger branches were infected and showed dieback symptoms.
- » From these branches the infection spread to the main stem often completely girdling and killing the upper part of the tree.
- » This was followed by complete defoliation and dieback of branches and death of the main stem in the upper crown.
- » It resulted in the development of epicormic shoots giving affected trees a bushy appearance with dead tops.
- » The infection progressed downwards killing all the primary and epicormic shoots.
- » The process of killing of the main stem, development and subsequent death of epicormic shoots was repeated until the infection had killed the entire tree.
- » A close examination of freshly affected trees showed that the infection was initiated in the top 25% of the crown.
- » In a typical outbreak of dieback, general decline in the growth rate of trees was observed.
- » Within two years, the foliage of the majority of trees previously healthy become progressively sickly and sparse in shoots and then branches dieback.
- » No infection of roots was observed.

Acacia auriculiformis

- » The infection occurred in the upper portion of the crown.
- » Initially, young shoots were affected and killed and the infection spread to the main stem.
- » As a result of this the main stem produced numerous epicormic shoots.

- As the infection moved downwards the epicormic shoots produced above died and fresh shoots were produced on the healthy part of the stem.
- This process continued until the infection had killed the main stem as well as all the side branches, eventually killing the whole tree; the roots remained unaffected.
- The main difference from the dieback of *A. mangium* was the greyish black colouration of the bark and sapwood, seen clearly after exposing it with a sharp knife.

Fungus: *Hendersonula* sp.

Management

- The control depended upon avoiding or minimizing stress caused by water shortage or extreme temperatures.
- Sanitary pruning was also practiced to eliminate the infected twigs and branches.

23. Root rot

Symptoms

- *Ganoderma* root rot is the most serious disease affecting commercially planted *Acacia mangium* in plantations in India and Indonesia, *Acacia catechu, A. arboretum, A. albida, A. aneura, A. decurrens, A. murrayana* and *A. victoriae* were the species most susceptible to the disease in India.

Acacia catechu

- Fructifications of the causal organism were found on the roots of dead and growing trees.
- The affected trees had dieback on upper crown.
- Roots were decayed and turned greyish black.
- Stumps colonized by the fungus were implicated as sources of infection.

Acacia mangium

- Typical symptoms of root rot were yellowing of phyllodes, premature defoliation and death of branches initially only in the upper crown and later throughout followed by epicormic shoots on the dying main stem of trees in patches.

- The affected trees showing different stages of symptoms were felled and studied for disease etiology.
- The stem remained unaffected and only the roots were affected with extensive browning.
- Fructifications of the causal organism growing at the base of dead roots of living trees were also observed.

Acacia auriculiformis

- The affected trees showed yellowing of phyllodes, thinning of crown due to premature defoliation followed by death of trees.
- The fructifications of the fungus were produced at the base of dead stems just after the first showers in June.
- On dead stems numerous fructifications developed during June-September.

Fungi: *Ganoderma philippii, Ganoderma mastoporum, Ganoderma aff. steyaertanum, Ganoderma australe, Ganoderma mastoporum, Ganoderma cupreum, Ganoderma sinense* and *Amauroderma rugosum*

Management:

- The only practicable control measure in forest stands could be removal of basidiocarp from trees followed by drenching with Bordeaux mixture 1%.

24. Heart rot

Symptoms

- In *Acacia mangium* plantations, many unidentified saprophytic wood decay fungi were observed fruiting on woody debris, dead branches and on living trees.
- In felled trees, rot was observed in the heart wood.
- The heart rot area was pale with dark zonation lines.
- Wounds were usually the entry points for the wood decay fungi.

Fungi: *Phellinus pachyphloeus, Trametes palustris, Tinctoporellus epimiltinus, Rigidoporus hypobrunneus, Schizophyllum commune, Fomes noxius, Polyporus alveolaris.*

Management:

- Pruning and singling operations should be carried out only on young and small-diameter branches followed by drenching with Bordeaux mixture 1%.

25. Powdery mildew

Symptoms

Acacia mangium

- White patches caused by powdery hyphae and spores were found on the leaves.
- These patches increased and coalesced, spreading to whole leaves.
- The upper surfaces of leaves become heavily infected with powdery hyphae.
- The infected leaves eventually resulted in early defoliation.

Acacia auriculiformis

- In *A. auriculiformis* powdery mildew was caused by *Erysiphe quercicola* in India.
- Phyllodes of various stages of maturity were affected with powdery mildew.
- This was characterised by the presence of white powdery masses on the upper surface.
- In severe cases, the infection also extended to the lower surface.
- Severely infected phyllodes showed curling, puckering and browning and abscissed prematurely.

Acacia aulacocarpa

- Early symptoms appear as discrete, cobweb-like to powdery white patches of hyphae and spores on the primary, juvenile, pinnate leaves.
- As infection progresses, these patches increase in size and coalesce to form bigger patches, spreading to the secondary leaves or phyllodes, sometimes resulting in defoliation.
- The upper surfaces of heavily infected foliage become coated with a mat of superficial hyphae and spores, with the foliage looking as though dusted with powder, hence the name 'powdery mildew'.

» The symptoms can sometimes be mistaken for road dust, especially on trees planted near dirt roads.

» The infection leaves yellow blotches on the surface of the foliage.

Fungi: *Oidium* sp., *Erysiphe acaciae, Erysiphe quercicola*

Management

» Application of sulphur dusting with a powder duster or application of fungicides such as Chlorothalonil 0.2% or Triademefon 0.2% or Maneb 0.2% or Zineb 0.1% gives effective control.

26. Black mildew

Symptoms

Acacia crassicarpa

» The fungus formed black spots on the surface of the phyllodes.

» Under heavy infestation, the entire leaf surface could be covered by the fungus.

» The scale insects and mealybugs excreted honeydew that provided a food source for fungal growth.

» The infection was more on the upper than the lower phyllode surface.

» Heavily infected phyllodes turned yellow and growth of seedlings stunted.

» However, on older trees, the black mildew did not cause any serious damage.

» In severe cases the phyllodes defoliated prematurely.

» Occasionally the incidence of black mould was so high that it covered almost >75 percent of the phyllode.

» Spherical fruiting bodies, perithecium developed among the hyphae.

» Ascospores were pigmented septate and could be observed in sac-shaped ascus.

Acacia auriculiformis

» The disease appeared in the form of superficial black irregular to round patches initially on the upper surface.

- The fungus forms black, radiate, velvety colonies on the surface of the phyllodes.
- In cases of heavy infestation, the entire phyllode surface may be covered by the fungus.
- The infection is usually more frequent on the upper than on the lower phyllode surface.
- Sometimes young stems and twigs can also be infected.
- Minute spherical fruiting bodies develop on the fungal thallus.
- Under warm and humid conditions these patches enlarged and coalesced to give rise to large patches.
- In severe cases the infection also extended to the lower surface.
- Severely infected phyllodes abscissed prematurely.

Fungus: *Meliola brisbanensis*

Management

- Black mildew could be controlled by spraying fungicides and insecticides to eliminate insects and bugs.

27. Pink disease

Symptoms

Acacia catechu

- Infection of young branches leads to killing and in old branches cankers are formed in barks with longitudinal cracking/ splitting.
- These cankers serve as infection courts for heart rot fungi which cause heart wood decay and thereby wood gets disintegrated.
- Infection on stem and branches leads to girdling and death of parts above.
- Lower areas show epicormic branches with results in plant growth retardation.
- Later gum exudes from cankers.

Acacia auriculiformis

- The typical symptoms were yellowing and wilting of phyllodes of shoots followed by defoliation.
- On the affected branches the fungus produced only the cobweb and pustule stages and no perfect or necator stages were observed.

Fungus: *Phanerochaete salmonicolor*

Management:

- Clonal propagation of resistant clones.
- Do not injure the trees or make wounds.
- Spray with Bordeaux mixture 1% and Tridemorph 0.2%.
- Spray with Carbendazim 0.1% at a weekly interval.

28. Seedling blight

Symptoms

- The seedlings show blighting accompanied with defoliation and die back in nurseries. Pathogen produces hyaline, cylindrical conidia with oil droplet.
- Under humid conditions the disease severity is more.

Fungus: *Colletotrichum gloeosporioides*

Management

- Spray with Carbendazim 0.1% or Copper oxy chloride 0.05% or Chlorothalonil 0.1% or Mancozeb 0.2%.

29. Damping-off

Symptoms

- Soil borne fungal disease in nursery systemic diseases.
- Causes pre emergence damping off where seeds rot before emergence.
- Causes post emergence damping off where seedlings rot after they emerge.

- Water soaked lesions and browning at seedlings at collar region, root decay.
- Death of seedlings in patches.
- Water logging aggravates disease incidence.

Fungus: *Rhizoctonia solani*

Oomycete: *Pythium* sp., *Phytophthora* sp.

Management

- Avoid water logging, prepare raised beds.
- Seed treatment with Captan 2 g/ kg seeds or *P. fluorescens* 10 g/ kg seed.
- Avoid wounding plants as these will allow entry of the fungus.
- Adequate spacing to avoid crowding and the formation of high humidity.
- Do not overwater as disease is more severe in very wet soils.
- Remove affected seedlings.
- Drench with *Pseudomonas fluorescens* at 2.5 kg/ 50 kg FYM to nursery bed.
- Drench with Copper oxy chloride at 0.25 % or Carbendazim at 0.1%.

30. Leaf web blight

Symptoms

- It first appears on leaves close to the ground, as water soaked silvery grey blotches.
- The infected adjoining leaves become joined together by the fungal hyphae if caught in a spider's web hence the name web blight.
- The fungal invasion of foliage is rapid over the wet leaf surface during rains.

Fungus: *Rhizoctonia solani*

Management

- Disposal of leaf litter by burning and segregation of diseased seedlings soon after such seedlings are spotted which prevents lateral spread of the disease.

- Raising of seedlings in poly bags instead of beds; keeping seedlings in lots of 250-300 seedlings instead of 1000 seedlings in nurseries.
- Foliar spray with Triademefon 0.2%.

31. Stem canker

Symptoms

- Cankers were dead areas of bark along stem axis.
- Symptoms appeared when the dieback occurred in the trees partially or completely girdled, causing crown dieback.
- The pathogen gained access to stem tissue through wounds and killed cambial tissue and sapwood but did not cause decay.
- The host produced callus tissue around the canker, and the margin of the canker cracks
- Infertile soils and climates caused more susceptible condition for canker diseases.
- Cankers on the main stem, reduced the vigor and such trees were liabled to wind breakage.
- Canker caused by *Nectria* sp. produced two-celled ascospores; hyaline, ovoid or ellipsoid.

Fungus: *Nectria* sp.

Management

- Sanitation, removal and/or burning of cankered limbs or the whole trees where possible, was often the only control measure.
- Spray with Captafol 0.2% or Bordeaux mixture 8:8:100 immediately after leaf fall helps reduce *Nectria* infections in trees

32. Basal stem canker and root decay

Symptoms

- It is reported in *A. auriculiformis* in India.
- The affected trees had a basal canker characterised by appressed bark with cracks.

» Under the bark the sapwood was discoloured and dead with black zonation lines.

» The roots on the side of the canker were dead and decayed with black zonation lines.

» In some trees, the dead bark and the underlying dead sapwood were attacked by termites.

Fungus: *Botryodiplodia theobromae*

Management

» Remove affected portion from the main plant

» Drench with *Pseudomonas fluorescens* at 2.5 kg/ 50 kg FYM to main field

» Drench with Copper oxy chloride 0.05% or Carbendazim 0.2%.

33. Branch canker

Symptoms

» The incidence of the disease was very high and it caused extensive mortality of branches ultimately killing the trees.

» The canker developed as a depressed area initially only on one side of the branch.

» Tissue in the depressed area died and developed black streaks where the fructifications formed.

» The canker gradually girdled the tissues thus killing the whole branch.

Fungus: *Cytospora* sp.

Management

» Drench with *Pseudomonas fluorescens* at 2.5 kg/ 50 kg FYM.

» Drench with Copper hydroxide 0.05% or Carbendazim 0.2%.

34. Shoot blight

Symptoms

» The apical bud and 2-3 young phyllodes appeared blighted as they turned blackish brown.

» The tip of other phyllodes just below the blighted ones showed tip necrosis.

Fungus: *Phomopsis* sp.

Management

» Remove affected shoots from the main plant and wound dressing with Bordeaux mixture 1%.

35. Stem decay

Symptoms

» It is reported in *Acacia mangium* in India.

» The affected trees had basal cankers, characterised by appressed bark on one side.

» When the bark was removed and a cut was made, the stem showed typical browning due to decay.

» When the decay advanced and girdled the stem completely the trees showed yellowing of phyllodes, defoliation and thinning of crown and ultimately death.

» Partially affected trees were infested by *Xyleborus*.

» It was not clear as to how the infection was initiated.

» The presence of the basal canker suggested that the infection took place near the ground level.

» Decay was also observed around the holes drilled by *Xyleborus*, indicating that possibly the infection started from the insect tunnels.

» Since roots could not be excavated the possibility of root infection could not be ruled out.

Fungus: *Gliocladium* sp.

Management

» Remove affected shoots from the main plant and wound dressing with Bordeaux mixture 1%.

36. Basal stem decay

Symptoms

» It is reported in *Acacia mangium* in India.

» The affected tree had a prominent decayed area near the ground level without any bark; roots remained unaffected.

» On exposure, it was found that the sapwood had become very soft and spongy with black and white mycelial growth.

» The crown showed dieback of shoots and thinning of crown.

Fungus: *Basidiomycete*; spongy rot/white rot.

Management

» Remove affected shoots from the main plant and wound dressing with Bordeaux mixture 1%.

37. Root decay and root rot

Symptoms

Acacia crassicarpa

» The affected trees showed yellowing of phyllodes and premature defoliation resulting in thinning of crown.

» On excavation of the root system it was found that some of the roots of such trees were either already dead or in the process of dying; dead roots showed zonation lines.

» No fructifications were observed on the affected roots.

Acacia mangium

» The affected trees had extensive dieback of shoots, especially in the upper crown, bushy growth of epicormic shoots appeared due to death of the main leading shoot.

» The shoots also died as drying up of the main stem gradually advanced downwards.

» Roots of such trees were dying and showed typical decay symptoms.

» Feeder roots were decayed and turned greyish black and even some of the primary and secondary roots had greyish-black decay leading to the base of the stem.

Fungi: *Macrophomina phaseolina, Armillaria mellea, Coriolopsis aspera* and *Phellinus* spp. on *A. auriculiformis* and *A. nilotica; Phellinus noxius* on *A. mangium.*

Management

» Soil drench with *Pseudomonas fluorescens* at 2.5 kg/ 50 kg FYM.

38. Rust

Symptoms

» A range of symptoms is produced by infection with *Uromycladium* spp. of which the most characteristic is the formation of reddish-brown, globose galls several centimetres in diameter which occur on stems and shoots of the trees.

» The surfaces of mature galls become covered with cinnamon-coloured spores.

» Inflorescences, phyllodes, fruits and shoot tips can also be infected with smaller galls causing gross malformation.

» When fresh, the galls are covered with powdery masses of spores.

» Older galls become reddish brown and eventually blackened and are commonly invaded by tunneling insects, sometimes resulting in misidentification as insect-induced galls.

» Susceptible trees may bear thousands of branch galls which affect the form of the tree due to repeated branching.

» Foliage terminal to the infections dies due to girdling of the shoot and severely infected trees may die.

» They are light brown when young and spore producing, becoming dull-brown with age.

Fungi: *Uromycladium tepperianum, Uromycladium acacia, Uromyces digitatus, Aecidium* sp., on exotic acacia and *Ravenelia tandonii* on *Acacia catechu.*

Management

» Spray with Triadimenol 0.2% or Propiconazole 0.1%.

39. Pod rot

Symptoms

- Dull purplish brown spots were seen on green pods, especially in the lower and middle parts of the crown.
- These spots coalesced to form large necrotic areas.
- The affected pods dried up and shed prematurely.
- Seeds inside the pods were found to be shrivelled and brown in colour.

Fungus: *Fusicoccum* sp.

Management

- Spray with Copper hydroxide 0.05% or Carbendazim 0.2% during pod formation stage.

40. Stem malformation/ Wilt/ Die-back

Symptoms

- The bark and the wood surrounding the cankers were discolored and had a black appearance due to the exudation of gum.
- The discolored wood typically had a streaked appearance, turning a uniform dark brown to dark blue color with age.
- Affected trees ranged from 8 to 12 months in age.

Fungi: *Ceratocystis manginecans* and *Ceratocystis acaciivora*

Management

- All diseased trees had been pruned 6-12 weeks earlier and lesions appeared to originate from these wounds, suggesting the involvement of a wound colonizing pathogen.

41. Bacterial leaf blight

Symptoms

- Disease symptoms apparent on 5-6 week-old seedlings and blight develop within 1-2 weeks.

- Early symptoms appear as small red streaks on tip, middle, or basal parts of the phyllodes.
- The strikes expand in length and width along the veins and later turn into brownish red color.
- They may then unite and dry, forming a blight syndrome.

Bacterium: *Xanthomonas campestris*

Management

- Spray with Streptocyclin 200 ppm or Oxy-tetracyclin 200 ppm.

42. Spike disease

Symptoms

- Symptoms include foliage dwarfing, shortened internodes, excessive branching, pale green or reddish leaves standing out stiffly from twigs, suppressed flowering and small fruit.
- The shape of phyllode near the shoot apex were transformed into needle like structures
- The affected trees were stunted in growth, this dwarf caused witches' broom on branches, leaves were small and fan-like arrangement of branches.
- Spiked trees usually die within 3-5 years of attack.

Bacterium: *Candidatus* Phytoplasma sp.

Management:

- The only practicable control measure in forest stands could be removal of diseased parts of trees.

43. Green algal spot

Symptoms

- It is widely reported in *Acacia crassicarpa* in India.
- Young lesions first appear on the phyllodes as yellow pinpoint spots, usually on the upper surface.

» These spots expand into velvety gray-green patches that become purplish-black with a central orange to rust-coloured tuft.

» Severely infected twigs become enlarged and stunted with sparse, withered leaves.

» As the alga can flourish only in humid conditions and on slow-growing tissues, the symptoms are usually most severe on old or suppressed trees.

Alga: *Cephaleuros virescens*

Management

» Application of copper fungicides and slaked lime, timed to coincide with the end of the rainy season.

Other Diseases

44. **Wilt and root rot**: *Fusarium solani* on *A. nilotica* and *A. auriculiformis, Fusarium oxysporum, Fusarium moniliforme* on *A. auriculiformis*

45. **Stem and butt rot**: *Inonotus rheades* on *A. aulococarpa*

46. **Stem rot**: *Laricifomes concavus* on *A. aulococarpa*

47. **Leaf spot**: *Nigrospora sphaerica* on *A. mangium, A. auriculiformis, A. aulococarpa*

48. **Dieback, gummosis, root death**: *Armillaria mellea* on *A. auriculiformis*

49. ***Phyllosticta* foliar blotch**: *Phyllosticta* sp.

50. **Septoria leaf spot**: *Septoria mortolensis* on *A. nilotica*

51. **Stem and branch canker**: *Lasiodiplodia theobromae, Dothiorella* spp, *Valsa* sp., *Nattrassia mangiferae, Macrovalsaria megalospora*

Selected References

Arentz F. 1993. Root rot in plantations of *Acacia auriculiformis* and *A. mangium*. In Impact of Diseases and Insect Pests in Tropical Forests. (Eds K.S.S. Nair and J.K. Sharma) Proceedings of IUFRO Symposium, 23-28 November 1993, Kerala Forest Research Institute, Peechi, India.

Bagchee K. 1945. Pathological notes. No.2. Wilt and die back on shisham (*Dalbergia sissoo*), babul (*Acacia nilotica*) and Khair (*A. catechu*) in the artificial generation under agriculture-cum-forestry management. Indian Forester 71, 20-24.

Bagchee K. 1958. Diseases of forest trees in India - causes, effects and methods of prevention. Indian Forester 84, 407-417.

Bakshi BK. 1957. Fungal diseases of khair (*Acacia catechu* Willd.) and their prevention. Indian Forester 83, 41-46.

Bakshi BK. 1957. Fungal diseases of khair (*Acacia catechu* Willd.) and their prevention. Indian Forester 83, 41–46.

Banard EL, Schroeder RA.1984. Anthracnose of Acacia in Florida. Proceedings of the Florida State Horticultural Society.

Boa E, Lenné J. 1994. Diseases of nitrogen fixing trees in developing countries: An Annotated List. Natural Resources Institute, ODA, Chatham, U.K., 82 p.

Browne FG. 1957. Pests and diseases of forest plantation trees: an annotated list of the principal species occurring in the british commonwealth. Clarendon Press, Oxford.

Corner EJH. 1983. Advances Polyporaceas I. Amauroderma and Ganoderma. Nova Hedwigia. Beiheft 75, 1-182.

Crist CR, Schoeneweiss DF. 1975. The influence of controlled stresses on susceptibility of european white birch stems to attack by *Botryosphaeria dothidea*. Phytopathology 65, 369-373.

CSIRO. 1982. Australian Acacias. Leaflet No. 0, Revised 1982. Canberra, CSIRO.

De, Guzman. 1977. Potentially Dangerous Diseases of Forest Trees in the Philippines. Biotrop special publication.

Dick M. 1985. *Uromycladium* rusts of Acacia. Forest Pathology in New Zealand No. 15, 8 p.

Florence EJM, Balasundaran M. 1991. Occurrence of pink disease of *Acacia auriculiformis* in Kerala. Indian Forester 117, 494-496.

Gibson IAS. 1975. Diseases of forest trees widely planted as exotics in the tropics and the Southern Hemisphere. Part I. Important Members of the Myrtaceae, Leguminosae, Verbenaceae and Melioceae. Commonwealth Mycological Institute, Kew, Surrey, England.

Hodges CS Jr, Gardner DE. 1984. Hawaiian forest fungi. IV. Rusts on endemic Acacia species, Mycologia 76, 332-349.

IUFRO. 1996. Impact of Diseases and Insect Pests in Tropical Forests. Proceedings of the IUFRO Symposium 23-26 Nov, 1993. Peechi, India.

Josiah SJ, Allen-Reid D.1991. Important Nursery Insects and Diseases in Haiti and their management. Proceedings of the first meeting of IUFRO working party, 22-30 Aug, 1990.

Kamal SRP. 1980. Fungi of Gorakhpur XIX. Pseudocercospora. Sydowia 33: 157–161.

Lee SS. 1997. Diseases of Some Tropical Acacias in Penisular Malaysia. Proceedings of an International Workshop. Sumatra. 1996.

Manion PD. 1981. Tree Disease Concepts. Prentice-Hall, Inc, Englewood Cliffs, New Jersey.

Mehm, Kyi Win. 1991. Trial Planting of *Acacia senegal* and *Acacia auriculiformis* in the Central Dry Zone of Myanmar. FRI.

Mehrotra MD. 1990. *Rhizoctonia solani*, a potentially dangerous pathogen of Khasi pine and hardwoods in forest nurseries. European Journal of Forest Pathology. 20, 329-338.

Mohanan C, Sharma JK. 1988. Diseases of exotic acacias in India. Journal of Tropical Forestry. 4, 357-361.

Mohanan C, Sharma JK. 1988. Diseases of exotic acacias in India. Journal of Tropical Forestry 4, 357–361.

Morris MJ. 1987. Biology of the Acacia gall rust *Uromycladium tepperianum*. Plant Pathology 36.

Old KM, Lee Su See, Sharma JK, Zi Qing Yuan. 2000. A manual of diseases of tropical acacias in Australia, South-East Asia and India. Center for International Forestry Research (CIFOR). https://doi.org/10.17528/cifor/000639.

Old KM. 2000. A manual of Diseases of Tropical Acacias in Australia, South-East Asia and India. Center for International Forestry Research (CIFOR) Indonesia.

Saw Win, Mya Win. 1996. Site preparations, species and spacing trial for pulpwood plantations. FRI.

Schmelzer K. 1969. Strawberry latent ringspot virus in Euonymous, Acacia, and Aesculus. Phytopathol. Z. 66, 1–24.

Sharma JK, Florence EJM. 1996. Fungal pathogens as a potential threat to tropical Acacias: A case study of India. (KFRI Research Report No: 113: 44p)

Sharma JK, Mohanan C, Florence EJM. 1985. Disease survey in nurseries and plantations of forest tree species grown in Kerala.KFRI Res. Rep. No.36, Kerala Forest Research Institute, Peechi, Kerala, India, 268 p.

Thite SV, Tapia MC, Kore BA, Camacho-Tapia M, Tovar-Pedraza JM. 2017. First Report of powdery mildew caused by *Erysiphe quercicola* on *Acacia auriculiformis* in India. Plant Dis 101:1825.

Tun Aung. 1994. Study on the Quality of Gum from *Acacia senegal.* FRI.

8 Diseases of *Aegle marmelos* (Wood Apple)

S. No.	Disease	Pathogen
1.	*Alternaria* leaf spot	*Alternaria alternata*
2.	Stalk end rot	*Fusarium solani, Fusarium semitectum* var. *majus*
3.	*Aspergillus* rot	*Aspergillus awamori*
4.	Leaf blight	*Fusarium moniliformae*
5.	Shell soft rot	*Syncephalastrum racemosum*
6.	Bacterial shot hole and fruit canker	*Xanthomonas citri* pv. *bilvae*

1. Alternaria leaf spot

Symptoms

» The symptoms initially observed as light brown to dark brown spots of indefinite size appearing on leaves with light brown or dark brown rings.

» Infected leaves further get blighted and fall down after drying and in extreme cases of disease entire infected plants become died.

» The disease was observed both in nursery and orchards but in nursery it observed quite severe than orchard.

» Affected leaves brighten and fall.

Fungus: *Alternaria alternata*

Management

- » Soil drenching with Copper oxychloride 0.2% or Carbendazim 0.1% or Mancozeb 0.2% or Propiconazole 0.1% or Difenconazole 0.1%.

2. Stalk end rot

Symptoms

- » Immature dropping of young fruits during rainy season is the main symptom of this disease.
- » The disease is characterized by softening of rind and underlaying pulp.
- » The affected rind turns dark brown at a later stage of infection.
- » Fruits neither shrivel nor lose their shape until press.
- » Infected fruits fall when there is heavy rain or strong wind.

Fungi: *Fusarium solani, Fusarium semitectum* var. *majus*

Management

- » Spray with Captofol 0.05% or Captan 0.1% or Oxycarboxin 0.1%.

***3. Aspergillus* rot**

Symptoms

- » The disease resulted into 100% loss of fruit kept in store house, since the infection was latent.
- » In severe cases the white outer skin of the shell become soft and rotten. Inner pulp produces black sooty growth of the fungus.

Fungus: *Aspergillus awamori*

Management

- » Spray with Wettable sulphur 0.2% or Mancozeb 0.2% or Chlorothalonil 0.1%.

4. Leaf blight

Symptoms

- Cottony growth of fungal mycelium was observed just beneath the hard shell.
- The fungus which encircled the whole fruit turned it soft and pulpy.

Fungus: *Fusarium moniliformae*

Management

- Spray with Copper oxychloride 0.2% or Mancozeb 0.2% twice at 15 days interval.

5. Shell soft rot

Symptoms

- The affected fruits rotted quickly and were found not fit for consumption as the entire fruit pulp became unpalatable.
- Lesions developed as rapidly expanding water-soaked light brown rot patches with dark brown margins and a gelatinous texture. Lesions measured ~6 to 8 cm in diameter.
- The rot developed quickly, and on the affected portion, the gelatinous layer was easily removable.
- Internally, the rot progressed into the pulp and was colonized by white to black fungal mycelium.
- The affected fruit produced an unpleasant odor typically associated with decay.

Fungus: *Syncephalastrum racemosum*

Management

- Spray with Benomyl 0.2% twice at 15 days interval.

6. Bacterial shot hole and fruit canker

Symptoms

- Symptoms of the disease appears as round, water soaked spots (0.5 mm) surrounded by clear hole which later on increase in size and form brown lesions with saucer like depression in the center, surrounded by oily raised margin.

- On fruits, the lesions are round, raised, water soaked 47 mm size and without any halo later on depression are formed in the center of spots surrounded by irregular, oily, raised margin which become corky, irregular and chocolate brown in colour.
- On twigs, the lesions are large (150 x 4 mm) and infected tissue ruptures in the center giving a rough corky appearance.
- Bacterial exudates are found on all infected part of plant.

Bacterium: *Xanthomonas citri* pv. *bilvae* (Syn: *Xanthomonas campestris* pv. *bilvae*)

Management

- Spray with Copper hydroxide 0.2% twice at 15 days interval.

Minor Diseases

7. ***Phoma* rot**: *Phoma* sp.
8. **Dry rot**: *Aspergillus niger, Aspergillus fumigatus, Aspergillus luchuensis*
9. **Leaf blight and leaf spot**: *Myrothecium roridum, Colletotrichum gloeosporioides*
10. **Powdery mildew**: *Oidium* sp., *Sphaerothcea fuliginea, Oidium tingitatanium*
11. **Leaf spot:** *Fusarium roseum, Cladosporium cladosporoides*
12. **Fruit rot**: *Fusarium solani, Phoma glomerata, Othia spiracae*

Selected References

Giri D, Banerjee K, Laha SK, Khatua DC. 1989. Some diseases of horticultural and field crops. Environment and Ecology 7, 821-825.

Madaan RL, Gupta PC. 1985. A leaf spot disease of *Aegle marmelos* L. caused by *Alternaria alternata* (Fr.) Keissler. Indian Journal of Plant Pathology 3, 239.

Maurya S, Kumar R, Kumari A, Choudhary JS. 2011. First report of *Alternaria* leaf blight in bael (*Aegle marmelos* (L.) Corr.) from eastern plateau and hill region of India. Journal of AgriSearch 3(4), 248-250.

Mishra P, Singh SP, Tiwari M, Singh S, Ajeet V. 2017. Management of alternaria leaf spot (*Aegle marmelos* Correa) of bael. Journal of Pharmacognosy and Phytochemistry. SP1: 918-920.

Misra AK, Garg N, Yadav KK. 2016. First report of shell soft rot of bael (*Aegle marmelos*) caused by *Syncephalastrum racemosum* in North India. Plant Disease. 100:1779-1779. https://doi.org/10.1094/PDIS-12-15-1475-PDN

Ngoc BTL, Vernière C, Jouen E, Ah-You N, Lefeuvre P, Chiroleu F, Gagnevin L, Pruvost O. 2010. Amplified fragment length polymorphism and multilocus sequence analysis-based genotypic relatedness among pathogenic variants of *Xanthomonas citri* pv. *citri* and *Xanthomonas campestris* pv. *bilvae*. International Journal of Systematic and Evolutionary Microbiology 60, 515-525.

9 Diseases of *Ailanthus excelsa* (Indian Tree of Heaven)

S. No.	Disease	Pathogen
1.	*Cercospora* leaf spot	*Cercospora glandulosa*
2.	Leaf blight	*Alternaria* sp.
3.	Parasitic plant	*Helicanthus elastica*
4.	Leaf web blight	*Rhizoctonia solani*
5.	Heart rot	*Navisporus floccosus*
6.	Powdery mildew	*Erysiphe quercicola*
7.	*Phoma* leaf spot	*Phoma medicaginis* var. *pinodella*

1. Cercospora leaf spot

Symptoms

» Small circular spot with a margin develops on old leaves.

» As the disease advances, the centre of spot become white, thin and drops off leaving shot holes.

» In severe infection defoliation also occurs.

Fungus: *Cercospora glandulosa*

Management

» Removal of infected plant debris.

- Spray with Carbendazim 0.1% or Copper oxychloride 0.25% or Chlorothalonil 0.1% or Mancozeb 0.2%.

2. Leaf blight

Symptoms

- A brown water soaked lesion develops on the leaf and spreads fast to occupy major portion of the leaf leading to complete drying of the lamina.
- The disease can easily identified by presence of concentric rings in the blighted area showing the speculation of the fungus.

Fungus: *Alternaria* sp.

Management

- Removal of infected plant debris.
- Spray with Carbendazim 0.1% or Copper oxychloride 0.25% or Chlorothalonil 0.1% or Mancozeb 0.2%.

3. Parasitic plant

Symptoms

- It is a partial stem parasite.
- *H. elastica* belongs to the family Loranthecea is a parasitic plant species infesting various crops including cotton.
- It is light green in colour and has thread like stem twining around the host plant.
- They are leafy and bear white coloured flowers which occur in cluster.
- As soon as the parasite gets in touch with the host stem, small structures like haustoria penetrate the host's tissue and derive nutrients and water from the host.
- This infestation kills the entire tree and spreads very rapidly to other trees through birds.

Parasite: *Helicanthus elastica*

Management

- Mechanical removal or pruning the affected branches can control.
- Regular monitoring, timely location of incidence and removal of infested braches of trees from the plantations as soon as it is noticed can prevent spread of this plant parasite.
- Severely infested trees should be uprooted and kept in isolation or even burnt in order to prevent further spread of the plant parasite to other trees.

4. Heart rot

Symptoms

- The first symptoms indicative of infection were discoloration of heartwood.
- It becomes dark brown, sapwood becomes brown. Incipient decay in heartwood was difficult to detect but wood colour was intermediate between that of healthy wood and decayed wood, i.e. it was usually of darker colour than normal heartwood.
- The cross section of infected *A. excelsa* stem shows a barrier zone line indicating that almost all of the heartwood was infected and only a little portion of healthy wood was left.
- The dark brown zone indicates that the wood was in advanced stage of decay with only remnants of fibers in hole.
- As the decay proceeds to the top of the tree, the fruiting bodies develop at butt region.
- It spread more in a vertical direction than in a horizontal plane from the place of infection zone showing an hole in the fallen tree trunk.

Fungus: *Navisporus floccosus*

Management

- Removal of butt rots fungi from the infected portion.
- Swabbing with Bordeaux mixture 1%.

5. Powdery mildew

Symptoms

- White powdery mycelium was present in distinct, irregular, scattered spots on the younger leaves and almost continuously present on lower portion of the mature leaves.
- The disease affected young leaves giving an unpleasant appearance to plant due to curled, bubbled and crinkled young leaves.

Fungus: *Erysiphe quercicola*

Management

- Provide adequate space between the trees.
- Spray with Triadimefon 0.2% or Wettable sulphur 0.1%.

6. Leaf spot and blight

Symptoms

- Leaf symptoms included small circular, brown, necrotic spots with a chlorotic halo.
- In severe infections, leaf spots coalesced and resulted in leaf blight.

Fungus: *Alternaria alternata*

Management

- Spray with Mancozeb 0.2% or Zineb 0.1%.

7. *Phoma* leaf spot

Symptoms

- The infection started from the middle part of the leaflets as small discoloured spots.
- Later, the spots become circular to irregular, light brown surrounded by reddish halo.
- The spots were restricted by midribs and veins.
- In severe stage the diseased spot separated out and form shot-hole.

Fungus: *Phoma medicaginis* var. *pinodella*

Management

» Spray with Carbendazim 0.1% or Chlorothalanil 0.1%.

Other Diseases

8. Wood decay: *Bjerkandera adusta*

***9. Corynespora* leaf spot**: *Corynespora cassiicola*

***10. Cercospora* leaf spot**: *Cercospora ailanthicola*

Selected References

Nagadesi PK, Arya A. 2014. A new Heart rot disease in *Ailanthus excelsa* Roxb. caused by *Navisporus floccosus* (Bres.) Ryvarden. International Letters of Natural Sciences. 6, 1-7. https://doi.org/10.18052/www.scipress.com/ILNS.6.1.

Nayak AK, Babu BK, Srivastava AK. 2019. Identification of *Erysiphe quercicola* associated with powdery mildew disease on *Ailanthus excelsa* in India. *Australasian Plant Pathol.* 48, 267–270. https://doi.org/10.1007/s13313-019-0626-8

Pramod S, Koyani RD, Bhatt I, Vasava AM, Rajput KS. 2015. Histological and ultrastructural alterations in the *Ailanthus excelsa* wood cell walls by *Bjerkandera adusta* (Willd.) P. Karst. International Biodeterioration Biodegradation, 100, 124–132

Rai M, Rajak RC. 1982. A new leaf spot disease of *Ailanthus excelsa* Roxb. Current Science. 51(2), 98-99.

10 Diseases of *Ailanthus triphysa* (Maharukh)

S. No.	Disease	Pathogen
1.	Damping off	*Pythium* sp.
2.	Collar rot	*Rhizoctonia solani*
3.	Seedling blight	*Colletotrichum dematium*
4.	Sooty mould	*Meliola ailanthii*
5.	Shot hole leaf disease	*Glomerella cingulata*
6.	Pink disease	*Phanerochaete salmonicolor*
7.	Stem canker	*Botryodiplodia theobromae*
8.	Bacterial leaf spot	*Ralstonia solanacearum*

1. Damping off

Symptoms

- Initially water-soaked lesions, measuring 5- 10 mm across, appeared on the hypocotyl near the ground level.
- The lesions turned brown in colour and the affected area got shrunken due to rapid collapse of cells, resulting into a prominent constriction.
- At this stage the seedlings fell over the ground and eventually died.

Oomycete: *Pythium* sp.

Management:

- The disease can be controlled by adopting silvicultural practices for avoiding excessive humidity in nursery beds.
- Soil drenching with Copper oxychloride 0.025% or Mancozeb 0.05% at monthly intervals.

2. Collar rot

Symptoms

- First symptom of the disease was the appearance of water-soaked lesions at the collar region.
- These lesions developed into light brown necrotic area and the tissue got decayed, which resulted in a constriction at the collar region and consequently death of seedlings.
- Young seedlings (I-month-old) usually collapsed from the decayed region and fell over the ground; old seedlings (> 2-month-old) showed only wilting.
- Occasionally infection of hypocotyl region in emerging seedlings was also observed.
- Affected seedlings either failed to emerge or emerged only partially, which died later.

Fungus: *Rhizoctonia solani* Kuhn state of *Thanatephorus cucumeris*

Management

- Soil drenching with Captan 0.025% or Carboxin 0.05% or Carbendazim 0.1%.

3. Seedling blight

Symptoms

- Initially the infection appeared on the stem near the apex in the form of elongated brown spots, which soon coalesced to form large necrotic area, Thereafter, the shoots showed typical symptoms of seedling blight i. e., wilting and drying up of leaves followed by death of the terminal bud.
- The necrosis extended downwards which eventually killed the seedling.
- During the monsoon, on the affected stem the pathogen produced numerous splash dispersed conidia in acervuli, which helped in the spread of infection.

Fungus: *Colletotrichum dematium*

Management

» Spray with Blue copper 0.02% or Thiram 0.02% or Carbendazim 0.1%.

4. Sooty mould

Symptoms

» The disease appeared in the form of superficial black irregular to round patches initially on the upper surface.

» Under warm and humid conditions these patches enlarged and coalesced to give rise to large patches.

» In severe cases the infection also extended to the lower surface. Severely infected leaves turned yellow and defoliated prematurely.

Fungus: *Meliola ailanthii*

Management

» Foliar spray with starch 5% followed by systematic insecticide.

5. Shot hole leaf disease

Symptoms

» Initially symptoms developed on the leaflets as dark green areas lined with a yellowish green margin.

» The colour of these areas changed to light yellowish to reddish orange and finally light brown.

» The necrotic areas of the leaf became thin and got detached very easily either by wind or due to impact of rain drops, thus leaving a prominent hole in the lamina.

» A severely infected leaflet got completely deformed with irregular cuts due to separation of a large part of the lamina after the infection.

Fungus: *Colletotrichum* state of *Glomerella cingulata*

Management

» Spray with Chlorothalanil 0.2% or Carbendazim 0.1%.

6. Pink disease

Symptoms

- The disease was characterised by a canker formed on the upper half of the stem, which resulted in the girdling of phloem thus killing the stem above the canker.
- The stem got constricted at the canker region and bark showed longitudinal splitting.

Fungus: *Phanerochaete salmonicolor*

Management

- Diseased tree can be managed by following proper hygiene practices.
- Spray with Carbendazim 0.1% at a weekly interval.

7. Stem canker and dieback

Symptoms

Canker

- The symptoms of stem canker is initially small longitudinal cracks appeared on the main stem from which occasionally a dark gummy substance oozed out.
- These cracks developed into large cankers, measuring 30-60 cm in length.
- The tissue under the canker turned greyish.
- Due to the infection numerous epicormic shoots developed just below the canker.
- The epicormic shoots also got infected in due course due to spreading canker, on which numerous fructifications of the causal organism produced.

Dieback

- The infection started near the ground at the root collar and rapidly spread upwards.
- Due to the infection, which was several millimetres deep, the, plants-showed decline symptoms with a weak crown having fewer and yellowed 1eaves.
- The weakened main shoot as well as the branches got infected and died slowly and gradually.

» As a result, numerous epicormic shoots were produced.

» In the meantime the basal stem canker had also spread downwards decaying the root system.

» Complete girdling of the outer bark killed the trees outright.

Fungus: *Botryodiplodia theobromae*

Management

» Diseased tree can be managed by following proper hygiene practices.

8. Bacterial leaf spot

Symptoms

» The first symptom of the disease was the appearance of water-soaked translucent amphigenous round to irregular lesions, which soon turned light brown.

» The lesions enlarged and coalesced to give rise larger irregular spots, which occasionally covered a considerable part of the lamina.

» The spots were shiny and sticky to touch and the necrotic tissue of the spots became thin and somewhat elastic.

» Severely infected leaflets were defoliated prematurely and seen sticking on other healthy leaves or stem.

Bacterium: *Ralstonia solanacearum*

Management

» Spray with copper based fungicides or Plantamycin 0.01%.

Selected References

Sharma JK, Mohanan C, Florence EJ. 1984, Diseases survey in nurseries and plantations of forest tree species grown in Kerala For. Res. Inst. Peechi, p.36.

11 Diseases of *Albizia lebbeck* (Black Siris)

S. No.	Disease	Pathogen
1.	Leaf spot and blight	*Cercospora albizziae, Colletotrichum* sp., *Alternaria alternata, Helminthosporium albizicolum, Camptomeris albizzia, Epicoccum* sp.
2.	Leaf web blight	*Rhizoctonia solani*
3.	Leaf rust	*Ravenalia clemensiae, Ravenalia sessilis, Sphaerophragmidium acaciae*
4.	*Fusarium* wilt	*Fusarium oxysporum* f. sp. *perniciosum, Fusarium solani*
5.	*Ganoderma* root rot	*Ganoderma lucidum*
6.	*Polyporus* root rot	*Polyporus shorea, Polyporus gilvus*
7.	*Botryodiplodia* die-back	*Botryodiplodia theobromae, Botryodiplodia hypodermia*
8.	Bacterial wilt	*Ralstonia solanacearum*
9.	Giant mistletoe	*Dendrophthoe falcata*
10.	Dodder	*Cuscuta reflexa*

1. Leaf spot and blight

Symptoms

» Small water soaked yellowish spot on leaves which later develops into a circular light brown lesion with a distinct yellowish margin or the disease appears first as minute brown dots, becoming cylindrical or oval; several spots coalesce to larger necrotic lesions and leaf dries up.

- Shot hole develops in the advanced stages of infection causing premature defoliation.

Fungi: *Cercospora albizziae, Colletotrichum* sp., *Alternaria alternata, Helminthosporium albizicolum, Camptomeris albizzia* and *Epicoccum* sp.

Management

- Collection and destruction of leaf debris.
- Spray with Captan 0.2% or Mancozeb 0.2% or Carbendazim 0.1% fungicide solution as foliar spray twice at 15 days interval.

2. Leaf web blight

Symptoms

- The disease appeared in seedbeds as irregular patches of web entangling seedlings.
- These patches enlarged rapidly from the periphery affecting the neighbouring healthy seedlings under high humidity and high seedling density.
- Occasionally, the disease covered the whole seedbed.
- The infected plants show water soaked grayish brown patches that enlarge rapidly and cover a large part or the entire lamina.
- Gradually the infection also spread to the rachis resulting in drooping of the whole leaf.
- Soon leaves 'turned brown and premature defoliation and abscission of the rachis occurred.
- The disease spread in a seedling from lower to upper whorl of leaves and from seedling to seedling through contact.
- The higher the seedling density to be greater was the spread of web blight in seedbed.
- The younger seedlings (1 - 2 month - old) were killed outright due to infection but in the older seedlings only defoliation was observed.
- In main field, shedding of infected tissues and formation of holes during heavy rains.

- The infected leaves dry up and are eventually shed.
- The disease spreads laterally in the nursery through overlapping foliage of the adjoining seedlings often resulting in group blighting of seedlings.
- In each case of severe infection, defoliation is high.

Fungus: *Rhizoctonia solani* imperfect state of *Thanatephorus cucumeris*

Management

- Remove and discard fully affected diseased plants.
- Spray with Mancozeb 0.2% twice at 15 days interval.
- Soil application of Carbendazim 0.1% or Oxycarboxin 0.1% around infected plants.

3. Leaf rust

Symptoms

- Foliar fungal disease in nursery.
- The pathogen attacks leaflets of the seedlings.
- Profuse development of orange yellow pustules on lower leaf surface.
- Later grayish fungal growth is observed in the infected area which represents teliospores.
- Finally leaves dry and defoliate.

Fungi: *Ravenalia clemensiae, Ravenalia sessilis, Sphaerophragmidium acacia*

Management

- Spray with Mancozeb 0.2% twice at 15 days interval.

4. Fusarium wilt

Symptoms

- Leaves become yellow stunted, and wilted affected branches may prematurely defoliate.

- Brown streaking in the wood, later cracks begins to appear in the bark.
- Gum or a white, frothy liquid exude from cracks.
- Fungus colonizes and clogs the tree's vascular (water conducting) tissue, results rapid tree death.
- After the tree dies, roots may still be alive and shoots may continue to sprout from the base of the trunk for some time.
- The roots of such seedlings showed prominent discoloration.

Fungi: *Fusarium oxysporum* f. sp. *perniciosum* on nursery and field of *Albizzia lebbeck, Albizzia julibrissin; Fusarium solani* on nursery of *Albizzia falcataria*

Management

- Remove affected branches, apply Bordeaux paste.
- Soil drenching with Carbendazim 0.2% or Mancozeb 0.2% or Thiram 0.1% or Chloroneb 3 g/ lit at monthly intervals.
- Soil application of organic manure + neem cake + *Trichoderma viride* (1 kg/tree).

***5. Ganoderma* root rot**

Symptoms

- Pathogen enters through distal root end.
- Pale leaves which later dries white mycelial strands seen in between bark and wood.
- When most of roots are affected, trees are killed.

Fungi: *Ganoderma lucidum* on *A. chinensis*, *A. procera* and *A. lebbek*; *Ganoderma pseudoferreum* on *A. lebbek; Ganoderma applanatum* on *A. procera*

Management

- Maintain the trees in healthy condition – application of manure.
- Do not wound the tree. If any wounds are there apply Bordeaux paste.
- Remove the sporocarp and destroy.

» Scrap the mycelia portion and apply Bordeaux paste.

» Remove severely affected trees.

» Isolate the affected trre from healthy trees.

» Drench the soil near the tree with 40 litres of Bordeaux mixture 1% at quarterly interval for thrice a year and repeat after 2-3 years.

6. Polyporus root rot or Patridge root rot

Symptoms

» Root gets affected at distal end and infection progresses up to collar region.

» Top drying of branches is observed which leads to death of trees.

» Trees become windblown, sometimes epicormic branches develop.

» Small yellow felt of mycelium of fungus is seen.

» Root injury or damage leads to infection, spreads by root contact and root grafting

» Heavy growth of weeds favours disease.

Fungi: *Polyporus shorea*, *Polyporus gilvus* on *A. procera*

Sporophores are formed at the base or on exposed roots when fresh, sessile, substipitate, funnel shaped, soft become brittle when dry, light in weight, single or imbricate, upper surface glabrous, brown to black and lower hymenium surface brown to black.

Management

» All dead, dry stumps and drying trees should be removed.

7. Botryodiplodia die-back

Symptoms

» The initial symptom of die-back was appearance of a stem canker in the form of a depressed greyish-black area, generally near the ground level, during the dry period.

» This was followed by yellowing of leaflets, which gradually defoliated prematurely.

- Slowly shoots in the upper crown of the tree showed symptoms of die-back.
- The bark over the canker splitted longitudinally and got separated.
- The wood of the affected trees showed greyish-black discolouration in streaks running vertically due to profuse mycelial growth in ray cells.
- As the canker advanced further, more branches died, including the main terminal shoot and the tree appeared to be almost dead.
- However, during the following monsoon numerous epicormic shoots developed from the living part of the stem.
- The infection progressed further upwards into stem and gave rise to stem canker.

Fungi: *Botryodiplodia theobromae*, *Botryodiplodia hypodermia.*

Associated fungi: *Hypoxylon rubiginosum* var. *tropicum* and *Hypoxylon bovei.*

Management

- All dead, dry stumps and drying trees should be removed.
- Spray with Carbendazim 0.2% or mancozeb 0.2% at monthly intervals.

8. Bacterial wilt

Symptoms

- In the affected plants, initially leaves of lower branches on one side turned yellow and finally wilted and dried up.
- Such plants had decayed feeder as well as primary roots which became greyish black.
- As the root infection proceeded further towards the stem, more side branches were killed.
- Finally, when the infection had already reached above ground, appearing as greyish black sunken canker on one side of the stem the terminal shoot got killed.
- From the healthy side of the stem numerous epicormic branches developed, which did not survive for long as the stem was completely girdled by the spreading canker down below at the base.

- When a small piece of stem/root taken from the affected trees was dipped in water streaks of bacterial ooze from the vascular elements were noticed.
- This confirmed the association of a bacterium with the disease.

Bacterium: *Ralstonia solanacearum*

Management

- Spray with Oxytetracyclin 500 ppm.

9. Flowering parasite (Giant mistletoe)

Symptoms

- The parasite is a partial parasite.
- It produces a very hard flattened prostrating stem which runs over the stem portions of the host and produce haustorial connections to drive the food source.
- The point of parasitism becomes thickened to form a gall like structure from which multiple shootings ranging from 15-20 arise as a clusters.
- The affected branch starts withering; drying from tip and dries off and produces erect, clustered branches from the galled portion.
- The parasite is a shrubby plant with green leaves.
- It produces cluster of orange coloured tubular flowers, produces dark red coloured berries.
- The birds are attracted by the fruits and act as dissemination agents.
- The seeds dropped on the crevices of the stem germinate and produce haustoria to attach to the host and proliferate in large numbers.

Parasite: *Dendrophthoe falcata*

Management

- Removal of parasite affected branches and apply Bordeaux paste at cut ends.
- Collect the seeds of parasite and destroy.

10. Flowering parasite (Dodder)

Symptoms

- Dodder identified by its thin stems appearing leafless / leaves reduced to minute scales.
- Dodder flowers are white to pink / yellow to cream.
- After a dodder attaches itself to a plant, it wraps itself around it.
- Produces haustoria that insert themselves into the vascular system of the host.
- The original root of the dodder in the soil then dies.
- The dodder attaches itself to multiple plants. May reach high into the canopy of shrubs and trees.

Parasite: *Cuscuta reflexa*

Management

- Remove affected portions at point of attachment and apply Bordeaux paste to cut end.

Other Diseases

11. **Leaf spot**: *Phomopsis mendax, Cercosporella theae* on *Albizzia lophantha, Periconia byssoides, Alternaria alternata, Epicoccum nigrum, Cercospora glauca, Cercospora albiziae, Colletotrichum dematium, Helminthosporium albiziicolum, Phyllosticta albizzinae* on *Albizzia lebbeck* and *Albizzia procera, Heterosporium albizziae* on *Albizzia julibrissin, Cladosporium albizziae, Colletotrichum lebbeck* on *Albizzia lebbeck* and *Albizzia procera*
12. **Yellow brown** spot: *Pleiochaeta albiziae*
13. **Leaf blight**: *Phytophthora nicotianae* and *Cladosporium* sp. on *Albizzia lebbeck*
14. **Leaf blotch and foliar necrosis**: *Camptomeris albiziicola* on *Albizzia lebbeck*
15. **Leaf fall**: *Phyllosticta albizinae, Endothella albizae* on *Albizzia lebbeck*
16. **Seedling blight**: *Colletotrichum capsici* on *Albizzia falcataria*
17. **Seedling wilt**: *Fusarium oxysporum* on *Albizzia procera*

18. **Rust**: *Haploravenelia japonica* on *Albizzia julibrissin, Sphaerophragmium acacia* on *Albizzia lebbeck, Ravenelia clemensiae* on *Albizzia procera*

19. **Powdery mildew**: *Oidium* spp

20. **Foliar necrosis**: *Camptorneris albizziae*

21. **Pink disease**: *Phanerochaete salmonicolor*

22. ***Aglaospora* root rot**: *Aglaospora* sp.

23. **Root rot**: *Armillariella mellea, Irpex subvinosus, Poria hypolateritia*

24. ***Fomes* stem canker/ Brown root rot**: *Fomes noxius*

25. **Black root rot**: *Macrophomina phaseolina*

26. **Purple root rot**: *Helicobasidium compactum*

27. **Violet root rot**: *Sphaerostilbe repens*

28. **Canker**: *Nectria pulcherrima, Nectria ditissima*

29. ***Macrophoma* stem infection**: *Macrophoma theicola*

30. ***Phoma* die-back**: *Phoma* sp.

31. **Charcoal stump rot**: *Ustulina zonata, Ustulina deusta.*

32. **Die-back**: *Physalospora rhodina, Thyridaria tarda, Scytalidium dimidiatum* on *Albizzia lebbeck*

33. **Canker**: *Fusicoccum* sp on *Albizzia julibrissin, Fusarium solani* on *Albizzia lebbeck* and *Albizzia procera*

34. **Needle cast**: *Phaeocryptopus gaeumannii* on *Albizzia julibrissin*

Selected References

Browne FG. 1968. Pests and Diseases of Forest Plantation Trees. Clarendon Press, Oxford. 1330 p.

Thomas WB, Jr. 1962. Reaction of biotypes of *Rhizoctonia solani* to different fungicides. Phytopathology 522: 366 (Abstr.)

Toole ER. 1966. Root rot caused by *Polyporus lucidus*. Plant Dis. Rep. 50: 945-946.

Wu L, Li SK. 1989. *Fusicoccum* canker, a new disease of Albizia julibrissin. Forest Pest and Disease. 2(34)

Mehrotra MD. 1989. *Rhizoctonia* leaf web blight of *Albizia l*ebbek, a destructive disease in forest nurseries in India. European Journal of Forest Pathology, 19(5-6), 382-384.

Bandara PD. 1987. Insects and diseases of forest plantations in Sri Lanka. In: Hitacharen C, MacDicken KG, Ivory MH, Nair KSS, eds. Proceedings of the IUFRO Workshop-Pests and Diseases of Forest Plantations. Regional Office for Asia and Pacific, FAO, Bangkok, 81-85

Harsh NSK. 1993. Fungicidal control of damping-off in seedlings of *Albizia lebbek*, *Dalbergia sissoo* and *Leucaena leucocephala* caused by *Fusarium* spp. Commonwealth Forestry Review, 72(1), 53-56

Ketaki C, Khan SN, Mehrotra MD. 1992. Occurrence of virus disease on *Albizia lebbeck*. Indian Journal of Forestry, 15(3):283-284

Sharma JK, Bhardwaj LN. 1988. Forest nursery diseases and their management. In: Ram Parkash, ed., Advances in Forestry in India, Vol. II, Debra Dun, India: International Book Distributors, 91-118

Gupta RK. 1993. Multipurpose trees for agroforestry and wasteland utilisation. Multipurpose trees for agroforestry and wasteland utilisation., xv + 562 pp.

Lenné JM, Boa ER. 1994. Diseases of tree legumes. Forage tree legumes in tropical agriculture., 292-308;

Gibson IAS, 1975. Diseases of forest trees widely planted as exotics in the tropics and southern hemisphere. Part 1. Important members of the Myrtaceae, Leguminosae, Verbenaceae and Meliaceae. Diseases of forest trees widely planted as exotics in the tropics and southern hemisphere. Part 1. Important members of the Myrtaceae, Leguminosae, Verbenaceae and Meliaceae., 51pp

Browne FG, 1968. Pests and diseases of forest plantation trees: an annotated list of the principal species occurring in the British Commonwealth. Clarendon Press, Oxford University Press, Oxford

Bakshi BK, 1976. Forest pathology: principles and practice in forestry. Forest pathology: principles and practice in forestry. Controller of Publications. New Delhi India, 400 pp

Seliskar CE, 1964. Virus and viruslike disorders of forest trees. Document, FAO/IUFRO Symposium on Internationally Dangerous Forest Diseases and Insects, Oxford No. FAO/FORPEST 64-5, 1966. pp. 44.

12 Diseases of *Aquilaria malaccensis* (Agarwood)

S. No.	Disease	Pathogen
1.	Damping-off	*Botrytis* spp., *Cylindrocladium* spp., *Fusarium* spp., *Rhizoctonia solani*, *Pythium* spp., *Phytophtora* spp.,
2.	Leaf spot	*Corynespora cassiicola*
3.	Wilt	*Fusarium* sp.
4.	Anthracnose and stem rot	*Colletotrichum gloeosporioides*
5.	Pink disease	*Phanerochaete salmonicolor*
6.	Die-back	*Lasiodiplodia theobromae*
7.	*Phytophthora* root rot	*Phytophthora parasitica*
8.	*Phyllosticta* leaf spot	*Phyllosticta capialensis*

1. Damping-off

Symptoms

» Poor seedling emergence with post- emergence, water-soaked stems, felled-over young plants leading to irregular, spreading patches of dead and dying plants.

» Young shoots and leaves show irregular stain-like patches followed by leaf decay.

» Rotting and dying root tips, visible spots along the stem, with sunken cankers.

» Browning or rotting of non-germinated seeds, which the stem of the seedling shrinks, rots near the stem collar, resulting to plant collapse.

Fungi: *Botrytis* spp., *Cylindrocladium* spp., *Fusarium* spp., *Rhizoctonia solani*

Oomycete: *Pythium* spp., *Phytophtora* spp.

Management

- » Biocontrol using arbuscular mycorrhizal fungus *Glomus fasciculatum* to control *Pythium* attack.
- » Prevention through use of pathogen-free seeds, media and water.
- » Raising pots one meter above ground.
- » Use of fungicide containing Thiabendazole 0.2% or Captan 0.1% or Benomyl 0.05% or Thiram 0.1%.

2. Leaf spot

Symptoms

- » The characteristic symptoms initially appeared on the upper surface of leaf as spindle shaped spot with yellow halo and pin head reddish brown dot at the centre.
- » Later on, the spots increased in size up to 2-3 mm.
- » The diseased spots were found scattered as dark brown to greyish black in colour, angular to oval and irregular shape, produced in large numbers on the surface of the leaves.
- » Initially symptoms were noticed on the younger leaves and later on infections spread to all the leaves of agar seedlings.
- » Immature leaf fall was also noticed in case of severe infection.

Fungus: *Corynespora cassiicola*

Management

- » Spray with Carbendazim + Mancozeb 0.1% or Chlorothalanil 0.1%.

3. Wilt

Symptoms

- » The leaves become wither suddenly.

- The leave become pale, yellowish and fall significantly.
- The shoot shrinks and dry.
- The disease will slowly attack the stem and root system.
- The color of Aquilaria sapling's bark becomes pale, brownish and darker.
- The side root become black, die and slowly decaying.

Fungus: *Fusarium* sp.

Management

- Maintain the cleanliness and control the moisture in nursery to prevent the propagation and spreading of fungus.
- Arrange and spacing the saplings to provide better ventilation and to prevent pathogen from spreading to adjacent polybag.
- Sowing / moving the saplings to higher location, at least 45 cm from ground level.
- Use a breed of seedlings that resistant to wilt disease.

4. Anthracnose and stem rot

Symptom

Anthracnose

- Black anthracnose on the tip of leaves.
- Fine black spots on the infected leaves.
- Anthracnose spread throughout the leaves.

Stem rot

- Fine fading spots / former petiole become black
- The disease will continue develop, the skin will sink and become soft
- The disease will continue develop if no action taken and will surround the stem
- No new shoot appear

- The leaves will wither, fade and fall
- The sapling will die

Fungus: *Colletotrichum gloeosporioides*

Management

- Destroy all infected leaves, trim and remove any dead branches on infected trees and adjacent to it.
- Provide sufficient fertilizer to ensure the tree always fertile and healthy.
- Control the hygiene, drainage, irrigation and ventilation in the nursery.
- Applying fungicides spray such as Thiabendazole 0.2% Thiram 0.1% or Mancozeb 0.1% or Chlorothalonil 0.2% or Thiophanate-methyl 0.1% or Propiconazole 0.1% before the infection occurs / become more serious.

5. Pink disease

Symptoms

- Initial infection can be seen by the presence of cob-web.
- Pink spots develop on the trunk skin.
- Later, pink crust was formed on the dead trunk.
- In severe infection, affected trees exhibit die-back symptoms with formation of elongated cracks on the trunk surface.
- The disease is spread by wind-borne spores that cause the existence of a pink fungus attached to the branches and trunks.
- White mycelia grow rapidly on the stem surface, followed by canker formation under infected bark.
- Formation of pink pustules.
- Epicormic shorts growing below the cankers.
- Wilting foliage on the infected stem, stem dieback, and breakage of the branches due to high wind.

- Pink disease often occurs in areas that receive high amounts of rain water or during rainy season. However, the spreading of the disease is also dependent on the healthy of trees and environmental conditions.

Fungus: *Phanerochaete salmonicolor*

Management

- Maintain the cleanliness and control the moisture in nursery to prevent the propagation and spreading of fungus.
- Arrange and spacing the saplings to provide better ventilation and to prevent pathogen from spreading to adjacent polybag.
- Sowing / moving the saplings to higher location, at least 45 cm from ground level.
- Use a breed of seedlings that resistant to wilt disease.

6. Die back

Symptoms

- Wilting trees with leaves attached on twigs and branches.
- Visible brownish to grey stain-like round spots occurring at lower part of the trunk, expands over time, then causes sunken stem surfaces, shrinkage, and cracks.
- Formation of galls and cankers on the stem.

Fungus: *Lasiodiplodia theobromae*

Management

- Maintain the cleanliness and control the moisture in nursery to prevent the propagation and spreading of fungus.

7. *Phytophthora* root rot

Symptoms

- Leaf wilting and premature drop-off.
- It initially starts off as a biotroph colonizing the root region, and then switches to become necrotrophic to cause tissue destruction and eventually root decay.

Oomycete: *Phytophthora parasitica*

Management

- Maintain the cleanliness and control the moisture in nursery to prevent the propagation and spreading of fungus.

8. *Phyllosticta* leaf spot

Symptoms

- Initially, necrotic, circular specks like spots were developed on the upper surface of leaf.
- Later, infected leaves are withering and premature drop-off was observed.

Fungus: *Phyllosticta capialensis*

Management

- Spray with Carbendazim 0.2% or Chlorothalanil 0.2%.

Selected References

- Borah RK, Ahmed FS, Sarmah GS, Gogoi B. 2012. A new record of leaf spot disease on *Aquilaria malaccensis* Lamk in India. Asian Journal of Plant Pathology, 6, 48-51.
- Das KK, Dubey LN. 2000. Studies of rhizosphere mycoflora of agar plants in Assam. Proceedings of the Seminar on Scope and Dimension of Agar Plantation in N.E Region; Assam. Hojai: All Assam traders and Agar Oil Manufacturers Association.
- Li Z, Chen J. 2008. Identification and characteristics of the damping-off pathogen on *Aquilaria sinensis* (Lour.) Gilg. Chinese Journal of Tropical Crops 29, 231-236.
- Premalatha K, Kalra A. 2013. Molecular phylogenetic identification of endophytic fungi isolated from resinous and healthy wood of *Aquilaria malaccensis*, a red listed and highly exploited medicinal tree. Fungal Ecology, 6, 205-211.
- Tabin T, Arunachalam A, Shrivastava K, Arunachalam K. 2009. Effect of arbuscular mycorrhizal fungi on damping-off disease in *Aquilaria agallocha* Roxb. seedlings. Journal of Tropical Ecology 50, 243-248.
- Zhang Z, Yang Y, Wei J, Chen X, Meng H, Wang M. 2013. Identification of pathogen of *Aquilaria sinensis* dieback disease. China Journal of Chinese Materia Medica 38, 1707-1711.

13 Diseases of *Azadirachta indica* (Neem)

S. No.	Disease	Pathogen
1.	Damping-off	*Fusarium* sp., *Sclerotium* sp., *Pythium* sp. and *Phytophthora* sp.
2.	Nursery root rot	*Rhizoctonia solani*
3.	Powdery mildew	*Oidium azadirachtae*
4.	*Cercospora* leaf spot	*Cercospora meliae*
5.	Leaf blight	*Alternaria alternata*
6.	Root rot	*Ganoderma lucidum*
7.	Seedling blight and root rot	*Fusarium solani*
8.	Web blight	*Rhizoctonia solani*
9.	Twig blight	*Colletotrichum gloeosporioides, Phoma jolyana, Phomopsis* sp.
10.	Bacterial blight and shot hole	*Pseudomonas azadirachtae*
11.	Giant mistletoe	*Dendrophthoe falcata*

1. Damping off

Symptoms

Pre-emergence damping off

» The young seedlings are killed even before they emerge from the soil.

- The fungus attacks the germinating seeds and they rot even before the hypocotyls emerge.
- Since this happens in the subsoil it cannot easily be spotted by the farmers who may have been misled on the quality of the seeds.

Post-emergence damping off

- This phase is characterized by toppling over of infected seedlings at any time between the period that they emerge from the soil and the stage the stem hardens to resist pathogen attack.
- Infection affects the stem at the ground level or spreads through the roots.
- The affected seedlings are pale green with a brown water soaked lesion at the base of the stem.
- Later, the lesion girdles the stem and spreads both upwards and downwards.
- The affected tissues are soft and water-soaked and they usually rot, leading to the collapse of the seedlings.
- The disease starts in patches and spreads to the entire lot in the nursery in the course of 2–4 days.

Fungi: *Fusarium* sp. and *Sclerotium* sp.

Oomycete: *Pythium* sp. and *Phytophthora* sp.

Management

- Soil solarization is used to kill resting structures.
- Avoid water stagnation.
- Formalin is applied as soil fumigant and the fungicides can be applied either as soil drench or soil mix.
- Seed dressing with fungicide Carbendazim at 0.2% seeds or *Trichoderma viride* 4 g/ kg seeds.
- Soil drenching with Carbendazim 0.2%.

2. Nursery root rot/ Web blight

Symptoms

- The pathogen is soil- borne and infects leaves of young seedlings.
- Greyish brown lesions develop on the leaf blade, advances and cover the entire leaf area.
- The fungal mycelia fully over the rotten tissues giving a spider web appearance and hence, the name web blight.
- The completely rotten tissues contain a large number of irregularly shaped sclerotia, the asexual resting structure which act as a source for the next crop.

Fungus: *Rhizoctonia solani*

Management

- Remove affected seedlings.
- Foliar spray and drench soil with Carbendazim 0.2%.

3. Powdery mildew

Symptoms

- White patches seen on the surface of the leaves.
- These patches coalesced and covered the whole leaf lamina giving greyish white appearance.
- Severely infected leaves and leaflets defoliated prematurely.

Fungus: *Oidium azadirachtae*

Management

- Foliar spray with Wettable sulphur 0.2%.

4. Cercospora leaf spot

Symptoms

- The infection starts as water soaked lesions, which later turns to minute grey spot with dark margins.

- » As the disease progresses, the spots become thin and papery and drops off leaving shot hole symptoms.

Fungi: *Cercospora meliae, Cercospora subsellis*

Management

- » Spray with Mancozeb 0.2% or Carbendazim 0.2%.

5. Leaf blight

Symptoms

- » A brown water soaked lesion develops on the leaf and spreads fast to occupy major portion of the leaf leading to complete drying of the lamina.
- » The disease can easily identified by presence of concentric rings in the blighted area showing the sporulation of the fungus.

Fungus: *Alternaria alternata*

Management

- » Spray with Mancozeb 0.2% or Copper oxychloride 0.2% twice at 15 days interval.

6. Root rot

Symptoms

- » The disease can be visualized as yellowing and dropping of leaves in one or two branches of the tress, subsequently followed by drying of the branches.
- » As the disease progresses, the defoliation and drying symptoms are seen on the other branches.
- » The fungus grows extensively from the bark and causes white rot of the sap wood which becomes spongy.
- » The dried trees show characteristic fruiting bodies of the fungus which arise from the base of the trees at the soil level.

Fungi: *Ganoderma lucidum, Ganoderma applantum*

Management

- Remove severely affected trees.
- Root feeding is also adopted where the root region is exposed; actively growing young root is selected and given a slanting cut at the tip. The root is inserted into a polythene bag containing fungicidal solution containing of Aureofungin sol 2g and copper sulphate 1 g in 100 ml of water is. The mouth of the bag is tied tightly with the root.
- Soil drenching with *Trichoderma viride* 200 g or *Pseudomonas maltophylla* 200 g or *Bacillus subtilis* 200 g with 50 kg of FYM.

7. Seedling blight and root rot

Symptoms

- The initial symptom water soaked region at the collar region.
- Later, they turn dark brown to black and progress upwards, girdling the collar region.
- Sudden drooping of leaves with withering of tips.

Fungus: *Fusarium solani*

Management

- Adopt raised bed nursery system.
- Soil drenching with neem cake 150 kg/ ha.
- Soil drenching with *Trichoderma viride* 200 g with 50 kg of FYM.
- Soil drenching with Copper oxychloride 0.2% twice at 15 days interval.

8. Web blight

Symptoms

- Initially appearance of greying brown blotches on leaves.
- The spots increase in size and coalesce together to cover the entire leaf blade.
- The infected adjoined leaves get joined together by the fungal hyphae as if caught in a spider's web.

- The diseased leaflets become detached prematurely.
- The detached leaflets remain clinging to the stem for a period of time by the fungal hyphae presenting a cob-web appearance.

Fungus: *Rhizoctonia solani*

Management

- Spray with *Trichoderma viride* 20 g / lit.
- Spray with Thiophanate methyl 0.2%.

9. Twig blight

Symptoms

- Initially appearance of small, circular to oblong, brown to black lesions on the succulent portion of the twigs or main shoot of seedlings, just below the growing tip.
- The lesion enlarged and girdled the stem, resulting into topple down of the seedling.

Fungi: *Colletotrichum gloeosporioides, Phoma jolyana, Phomopsis* sp.

Management

- Spray with Captofol 0.2% or Carbendazim 0.2%.

10. Bacterial blight and shot hole

Symptoms

- Three distinct types of symptoms were observed under natural infection: angular leaf spot, shot-hole and vein blight.
- Among meteorological factors, temperature was found to be the most important in influencing the symptom manifestation.
- Angular leaf spot first appears as small water-soaked translucent spots well delimited by veins and veinlets.
- Angular leaf spots in the later stages turn brown and necrotic, and subsequently fall off leaving circular or irregular perforations on the leaflet.

- The layer separating the holonecrotic area from the necrotic zone develops at a distance from the spot.
- These spots enlarge, coalesce and form bigger spots under favourable conditions.
- Shot-hole was a very characteristic feature of the disease in conditions favouring maximum disease development.
- Vein blight is characterised by a narrow translucent water-soaked area of infection 0.5-3 mm on either side of the midrib.
- It starts either from the top of the leaflet and extends towards the base, or from the base of the leaflet.
- Sometimes only the mid-portion is affected. The midrib turns brown, then black and finally collapses.
- The area adjoining the streak becomes chlorotic and subsequently the entire leaflet turns chlorotic.
- It subsequently dries and falls prematurely.

Bacterium: *Pseudomonas azadirachtae*

Management

- Spray with Streptomycin sulphate 200 ppm.

11. Flowering parasite (Giant mistletoe)

Symptoms

- The parasite is a partial parasite which has a very hard flattened prostrating stem which run over the stem portions of the host and produce haustorial connections to drive the food source.
- The point of parasitism becomes thickened to form a gall like structure from which multiple shootings ranging from 15-20 arise as a clusters. In the case of the severe parasitism, the affected branch start withering drying from tip and dry off.
- The parasite has dark green leaves and produce erect, clustered branches from the galled portion.

» It produces cluster of orange coloured tubular flowers.

» The parasite produces dark red coloured berries.

» The birds are attracted by the fruits and act as dissemination agents.

» The seeds dropped on the crevices of the stem germinate and produce haustoria to attach to the host and proliferate in large numbers.

Parasite: *Dendrophthoe falcata*

Management

» Removal of parasite affected branches and apply Bordeaux paste at cut ends.

» Collect the seeds of parasite and destroy.

Other Diseases

12. **Vascular wilt**: *Fusarium solani*
13. **Pink disease**: *Phanerochaete salmonicolor*
14. **Canker**: *Diatrype azedarachae*
15. ***Septoria* leaf spot**: *Septoria meliae*
16. ***Phyllosticta* leaf spot**: *Phyllosticta azadirachis*
17. ***Cercospora* leaf spot**: *Cercospora meliae*
18. ***Colletotrichum* leaf spot/ blight**: *Colletotrichum capsici, Colletotrichum gloeosporioides, Colletotrichum subsessilis*
19. ***Pseudocercospora* leaf spot/ blight**: *Pseudocercospora subsessilis*
20. ***Periconia* leaf spot**: *Periconia parasitica, Periconia byssoides*
21. **Leaf spots**: *Sclerotium rolfsii, Alternaria* sp., *Curvularia* sp., *Exserohilum* sp.
22. **Weeping canker and die-back:** *Botryodiplodia theobromae, Fusarium compactum*
23. ***Alternaria* leaf blight**: *Alternaria alternata*
24. **Leaf spot**: *Xanthomonas azadirachtii*

Selected References

Bakshi BK. 1976. Forest pathology principles and practices in forestry, Govt. of India Press, FRI and colleges, Dehradun, 400p.

Desai SG, Gandhi AB, Patel MK, Kotasthane WV. 1966. A new bacterial leaf spot and leaf-spotting fungi. Phytopathology, 18: 717-751.

Harsh NSK, Gupta BN. 1993. Root diseases in forest nurseries of Madhya Pradesh and their control. TFRI Infor Series 1. 31p.

Hegde NG 1993. Neem production and development constraints at grass root level. In: Souvenier of World Neem Conference, Indian Society of Tobacco Science, CTRI, Rajamundry (A.P.), India. 1-7p.

Jamaluddin, Harsh NSK Nath V. 1997. Hand book of diseases in tropical tree species TFRI. Pub 8.53 p.

Jamaluddin, Verma RK, Nath V. 1994. Diseases occurring in different provenances of neem in central India and their control. Indian J. Trop. Biodiv, 2: 466-468p.

Mridha MAU, Bhuiyan MK, Akhter MF, Mahmud M, Dhar PP, Jabbar F. 2005. Powdery mildew of *Azadirachta indica* A. Juss. caused by *Oidium azadirachtae* sp. nov. recorded in Bangladesh. Hamdard Medicus 48(4): 130-132.

Nagendra Prasad MN, Shankara Bhat S, Haraprasad N, Sreenivasa MY, Raveesha KA, Janardhana GR 2010. Study of die back disease incidence of neem in Karnataka, India and PCR based identification of the isolates, Archives of Phytopathology and Plant Protection, 435, 446-453.

Sharma JK, Mohanan C, Florence M. 1985. Disease survey in nurseries and plantations of forest tree species grown in Kerala, KFRI Res. Rpt. 36. 268p.

Soni KK, Pyasi A, Tiwari, P, Shirin F, Verma RK. 2000. Incidence of vascular wilt disease in different provenances of neem - (*Azadirachta indica* A. Juss.) in Central India and its Management. The Indian Forester, 139:12.

Srivastava SK, Patel PN. 1969. Epidemiology of bacterial leaf spot, blight and shot-hole disease of neem in Rajasthan. Indian Phytopath., 22: 237-244.

14 Diseases of *Bauhinia variegata* (Bauhinia)

1. Cladosporium leaf spot

Symptoms

- The infection starts as a small round or oval reddish brown coloured amphigenous spot which enlarges to irregular patches.
- Sometimes the smaller spots coalesce to form larger spots.
- Infected leaves often become chlorotic when a major portion of the leaf is infected; they fall off from the plant.

Fungus: *Cladosporium cladosporioides*

Management

- Removal of infected plant parts.
- Spray with Mancozeb at 0.25% or Copper oxychloride at 0.25%.

Other Diseases

2. **Seedling blight**: *Bartalinia robillardoides*
3. **Leaf tip blight**: *Glomerella cingulata*
4. **Leaf spot**: *Schiffnerula cassiae*

5. **Leaf spot**: *Cercospora bauhiniae, Phyllosticta bauhiniae*

6. **Inflorescence rot**: *Stigmina phaeocarpi*

Selected References

Jamaluddin, Harsh NSK, Nath V. 1997. Hand book of diseases in tropical tree species TFRI. Pub 8, 53 p.

15 Diseases of *Bombax ceiba* (Silk Cotton)

S. No.	Disease	Pathogen
1.	Brown leaf spot	*Pseudocercospora italica*
2.	Collar rot	*Rhizoctonia solani*
3.	Leaf blight	*Corticium rolfsii*
4.	Leaf rust	*Uredo bombacis*
5.	*Myrothecium* leaf spot	*Myrothecium roridum*
6.	*Colletotrichum* leaf spot	*Colletotrichum gloeosporioides*
7.	*Cercospora* leaf spot	*Cercospora bombycina*
8.	Pink disease	*Phanerochaete salmonicolor*

1. Brown leaf spot

Symptoms

» At first, small, round to irregular and yellowish spots, 2-5 mm in diam., appear at the central or marginal area of leaves.

» They soon enlarge to 5-10 mm in size and sometimes coalesce to each other to form bigger lesions.

» Then brown necrotic area arises at the center of yellowish spots.

» Lower surface of the necrotic spots is covered with dark greenish to blackish sooty masses consisting of conidiophores and conidia.

- The disease is usually prevalent on kapok seedlings and planted young trees.
- Diseased leaves easily defoliate by shaking them.
- In case of severely attacked seedlings, only a few leaves survive at the top of shoot or twigs and growth of seedlings is remarkably retarded.

Fungus: *Pseudocercospora italica*

Management

- Spray with Mancozeb 0.2% or Chlorothanil 0.2%.

2. Collar rot

Symptoms

- The initial symptom of the disease was the appearance of water soaked lesions on the hypocotyle.
- Later these lesions turned brown in colour and the infected area got decayed.
- Affected seedlings usually collapsed from the decayed portion, fell over the ground and died.

Fungus: *Rhizoctonia solani* Kuhn state of *Thanarephorus cucumeris*

Management

- Since the disease is manifested under high soil moisture regimes, the watering frequency.
- Population per bed should be reduced after the appearance of the disease to check its further spread.
- Drenching with Copper oxychloride 0.25% or Emison 0.005%.

3. Leaf blight

Symptoms

- The initial symptoms of the disease were the appearance of small circular brownish-yellow spots in concentric rings on the leaflets.

- These spots increased in size and coalesced to form large necrotic areas, which often covered the entire leaf and even the petioles.
- Soon, because of rotting of petioles, the leaves bent downwards and dried up.
- The infection spread rapidly through contact of diseased leaves with healthy ones causing extensive premature defoliation.
- Contact with stem caused stem decay which resulted in death of seedlings.
- On the affected leaves and stem numerous off-white sclerotia developed which fell over the ground after getting detached easily from the mycelium.

Fungus: *Corticium rolfsii*

Management

- Two foliar sprays of Emisan-6 0.005% at a weekly interval.

4. Leaf rust

Symptoms

- The rust infection first appeared only on leaflets of mature leaves.
- Later, as the top young leaves matured, the infection proceeded upwards and all the leaves got infected.
- Initially infection appeared as light greenish-yellow flecks on the upper surface during the rains.
- Uredinia developed in the form of light yellowish-orange blisters, 1-2 mm across, scattered on the lower surface.
- Later, after the eruption, the pustules appeared bright-orange-brown in colour.
- Upper surface of leaf showed characteristic yellowish-orange flecking at the location of uredinia on the lower surface.
- When the rust infection was severe (>25 uredinia cm-2) the whole leaflet turned yellow and defoliated prematurely.

Fungus: *Uredo bombacis*

Management

- » Spray with Oxycarboxin 0.1% at a weekly interval.

5. Myrothecium leaf spot

Symptoms

- » Initially, amphigenous, minute, circular light brown spots with dark brown margin developed on leaflets of any leaf irrespective of maturity but more on older ones.
- » Sporodochia developed on both the surfaces, usually more on the lower surface, at the periphery of spots.
- » The sporodochia produced abundant olive-green conidial mass which later turned dark in colour.
- » The fungus invaded the adjoining green tissues and formed large necrotic brown areas.
- » As the spots enlarged, more concentric rings of sporodochia were added.
- » The sporodochia of the inner rings became more or less inactive producing fewer conidia as compared to those situated in the outer ones.
- » Thus, each leaflet developed one to four large spots, circular, to oval with uniform to irregular dark brown margin, 15-25 mm in diam, which eventually coalesced to give rise to large necrotic area.
- » The severely affected leaflets dried up and defoliated prematurely.

Fungus: *Myrothecium roridum*

Management

- » Spray with Carbendazim + Mancozeb 0.1% at a weekly interval.

6. Colletotrichum leaf spot

Symptoms

- » The disease was characterized by circular, 1-2 mm diam., dark black leaf spots with a narrow yellowish green border on the upper surface of leaflets.
- » On the lower surface the spots appeared dull grey in colour.

- Under high humid conditions, occasionally conidia were produced on the under surface of the spots.
- The necrotic area of the old spots got detached and formed shot-holes.

Fungus: *Colletotrichum gloeosporioides*

Management

- Spray with Carbendazim 0.1% or Mancozeb 0.1% or Chlorothalanil 0.1% at a weekly interval.

7. *Cercospora* leaf spot

Symptoms

- The infection occurred on young as well as mature leaves as both were found to be equally susceptible to disease.
- Initially small pale brown spots appeared on the leaflets which turned into cream coloured necrotic areas with dark brown margins and light pale halos.
- Severe foliage infection leads to defoliation.

Fungus: *Cercospora bombycina*

Management

- Spray with Carbendazim + Mancozeb 0.1% at a weekly interval.

8. Pink disease

Symptoms

- The infection, characterised by pink encrustation over a canker, occurred somewhere on the upper half part of the stem.
- The stem above the canker rarely died due to incomplete girdling of phloem.

Fungus: *Phanerochaete salmonicolor*

Management

- Spray with Carbendazim 0.1% at a weekly interval.

Other Diseases

9. **Dieback**: *Colletotrichum capsici, Fusarium solani* and *Lasiodiplodia theobromae*

10. **Leaf spots:** *Cercospora ceibae*, *Corynespora cassicola, Phyllosticta bombacis*, *Phoma holoptelea*

11. **Nursery leaf spot:** *Colletotrichum gloeosporioides*

12. **Leaf bight:** *Ascochyta bombacina*

13. **Seedling blight:** *Sclerotium rolfsii* and *Ascochyta* sp.

14. **Anthracnose**: *Colletotrichum dematium*

15. **Sooty mould:** *Capnodium brazilienze*

16. **Collar rot**: *Rhizoctonia solani*

17. **Root rot**: *Armillaria ostoyae, Calonectria* sp. and *Camillea* sp.

18. **Tumour**: *Agrobacterium tumefaciens*

19. **Wood stains:** *Botryodiplodia theobromae, Ceratocystis fimbriata, Acremonium* sp., *Scytalidium lignicola* and *Fusarium* sp.

20. **Phanerogamic parasite:** *Dendrophthoe falcata*

Selected References

Parthasarathy S, Thiribhuvanamala G, Prabakar K. 2019. Diseases of Field Crops and their Management. Jaya Publications, New Delhi.

Sharma JK, Mohanan C, Florence EJM. 1985. Disease survey in nurseries and plantations of forest tree species grown in Kerala.KFRI Res. Rep. No.36, Kerala Forest Research Institute, Peechi, Kerala, India, 268 p.

16 Diseases of *Callistemon viminalis* (Bottlebrushes)

1. ***Cylindrocladium* leaf spot**

Symptoms

» Minute, Brown, pinpoint leaf spots (< 1 mm across) on both leaf surfaces, sunken blotches, and blight.

Fungi: *Cylindrocladium colhounii, Cylindrocladium clavatum*

Management

» Destruction of infected leaf debris.

» Spray of thiophanate methyl 0.1%, iprodione 0.2%, or mancozeb 0.1% based on the severity of the disease.

Selected References

El-Gholl, Schubert TS, Walker S.E. 1998. Cylindrocladium Leaf Spot of Callistemon. Plant Pathology Circular No. 389.

Horst RK. 2013. Westcott's Plant Disease Handbook. Springer Netherlands, ISBN: 978-94-007-2140-1

17 Diseases of *Calophyllum inophyllum* (Alexandrian laurel)

1. Takamaka wilt

Symptoms

- » The leaves lose their shine, curl inwards and dry out, and turn a bronze colour.
- » The leaves dry suddenly and can remain attached to the trees for up to 3 weeks.
- » No apparent lesions on the trunks, branches or roots are associated with the disease.
- » Most branches dead, only one branch or less with foliage.
- » Defoliation foliage, and extensive bark loss from major stems.

Fungus: *Leptographium calophylli*

Management

- » All infected trees were cut down and burned.
- » Soil drenching with organic amendments.

Selected References

Horst RK. 2013. Westcott's Plant Disease Handbook. Springer Netherlands, ISBN: 978-94-007-2140-1.

Hill M, Currie D, Shah NJ. 2003. The impacts of vascular wilt disease of the takamaka tree *Calophyllum inophyllum* on conservation value of islands in the granite Seychelles. Biodiversity and Conservation, 12(3), 555-566

18 Diseases of *Casuarina equisetifolia* (Beef wood)

S. No.	Disease	Pathogen
1.	Damping-off	*Pythium* spp.
2.	Leaf spot/ Needle blight	*Phoma casuarinae*
3.	Stem wilt/ Blister bark	*Trichosporium vesiculosum*
4.	Pink disease	*Phanerochaete salmonicolor*
5.	*Ganoderma* root rot	*Ganoderma lucidum*
6.	*Polyporus* root rot	*Polyporus gilvus*
7.	Bacterial wilt	*Ralstonia solanacearum*
8.	Flowering parasite (Giant Mistletoe)	*Dendrophthoe falcata*
9.	Flowering parasite (Dodder)	*Cuscuta reflexa*

1. Damping off

Symptoms

» Soil borne systemic fungal disease in nursery.

» Causes pre emergence damping off where seeds rot before emergence.

» Causes post emergence damping off where seedlings rot after they emerge water soaked lesions and browning at seedlings at collar region, root decay.

» Death of seedlings in patches.

» Water logging aggravates disease incidence.

Oomycetes: *Pythium debaryanum*, *Phytophthora cambivora, Phytophthora nicotianae* var. *parasitica*

Fungi: *Rhizoctonia solani*, *Fusarium lacertarum* and *Fusarium oxysporum*

Management

» Avoid water logging, prepare raised beds.

» Use of clean, healthy, and sterile seeds.

» Utilize pest-free soil or growing medium through incorporation of compost, plant residues, or microbial amendments into soil.

» Use relatively acidic soils with a low pH (4.5–6.0), increase soil pH with organic amendments, with applications of aluminum sulfate, sulfur, or acid peat.

» Avoid warm or wet weather for sowing.

» Avoid over-sowing or excessive plant densities, use crop varieties with asynchronous germination.

» Apply well-balanced fertilization especially microelements (phosphorus, potassium, and calcium).

» Avoid monoculture and adopt long rotation schemes to lower down pathogen populations, introduce *Brassica* crops.

» Perform tillage to incorporate plant residues into the soil to reduce soil-borne pathogen populations.

» Remove affected seedlings.

» Seed treatment with Captan 2 g/ kg seeds *P. fluorescens* at 10 g/ kg seed

» Avoid wounding plants as these will allow entry of the fungus.

» Adequate spacing to avoid crowding and the formation of high humidity.

» Do not overwater as disease is more severe in very wet soils.

» Apply *Pseudomonas fluorescens* or *Bacillus subtilis* or *Streptomyces* sp. at 2.5 kg/ 50 kg FYM to nursery bed.

» Drench with Copper oxychloride at 0.05% or Carbendazim at 0.1%.

2. Leaf spot/ Needle blight

Symptoms

» The disease develop as yellow discoloration from the tip of the needle initially and later spread to occupy major portion of the needle leading to blighting and drying of the needles.

» Later, the affected needles turn to black coloured due to production of hundreds of pycnidia, the asexual fruiting bodies.

Fungi: *Phoma casuarinae, Dasyschypha eucalypti*

Management

» Remove infected plant parts.

» Foliar spray with Mancozeb at 0.25% or Copper oxychloride at 0.25%.

3. Stem wilt/ Blister bark

Symptoms

» Discolouration of the foliage is the initial symptoms of blister bark disease.

» The main field symptoms are paling and drying of needles, branches and the main stem, followed by extensive production of black sooty conidia the bark is pushed outward that result in the formation of 'blisters'.

» Such blisters, of varying size and shape, can be located all over the main stem and branches of affected trees.

» The blisters eventually rupture in the form of scales or flakes and expose the powdery black mass of spores produces numerous spores that are carried by wind to healthy stems and enter through injury and cause primary infection.

» Later the pathogen travel up the stem and down to the roots and sporulates on roots, as a result disease occurs in patches.

» In severe cases, diseased trees develop vertical cracks and splits in the main stem and die.

Fungus: *Trichosporium vesiculosum*, reclassified into *Subramanianospora vesiculosa*

Management

- Selection and propagation of disease resistant germplasm is the only long-term management solution.
- When infected trees are spotted in plantations, stop pruning as the spores may enter through the branches.
- Isolate diseased tree by forming trenches.
- Avoid injury.
- Remove the infected trees immediately.
- Scrap the infected portion and spray with Copper oxychloride 0.05%, soil drench with Carbendazim 0.1%.
- Apply *Pseudomonas fluorescens* or *Trichoderma viride* 100 g/ tree to induce immunity.

4. Pink disease

Symptoms

- It is a fungal soil borne disease that usually occurs in young plantations (1-5 year) during monsoon period.
- Infection of young branches leads to killing and in old branches cankers are formed in barks with longitudinal cracking/ splitting.
- These cankers serve as infection courts for heart rot fungi which cause heart wood decay and thereby wood gets disintegrated.
- Infection on stem and branches leads to girdling and death of parts above.
- Lower areas show epicormic branches with results in plant growth retardation.
- Later gum exudes from cankers.

Fungus: *Phanerochaete salmonicolor* **(Syn:** *Corticium salmonicolor***)**

Management:

- Remove affected portions and apply Bordeaux paste at cut ends.

» Clonal propagation of resistant clones.

» Do not injure the trees or make wounds.

» Apply Bordeaux mixture 1% and Tridemorph 0.1%.

5. *Ganoderma* root rot

Symptoms

» Pathogen enters through distal root end.

» Pale leaves which later dry with white mycelial strands seen in between bark and wood, sporophores are seen at base of tree trunk, on stumps.

» When most of roots are affected, trees are killed.

Fungus: *Ganoderma lucidum*

Management

» Maintain the trees in healthy condition by the application of manures.

» Do not wound the tree. If any wounds are there apply Bordeaux paste.

» Remove the sporocarp and destroy.

» Scrap the mycelia portion and apply Bordeaux paste.

» Remove severely affected trees.

» Isolate the affected tree from healthy trees.

» Drench the soil near the tree with 40 litres of Bordeaux mixture 1% at quarterly interval for thrice a year and repeat after 2-3 years.

» Root injection with 2 g of Aureofungin soil and 1 g of copper sulphate in 100 ml of water is taken in a saline bottle and the bottle is tied with the tree.

6. *Polyporus* root rot (patridge root rot or sapwood rot)

Symptoms

» The disease is common in high rainfall areas, less in drier conditions.

- Root gets affected at distal end and infection progresses up to collar region.
- Top drying of branches is observed which leads to death of trees.
- Small yellow felt of mycelium of fungus is seen.
- Sporophores are formed at the base or on exposed roots when fresh, sessile, substipitate, funnel shaped, soft become brittle when dry, light in weight, single or imbricate, upper surface glabrous, brown to black and lower hymenium surface brown to black.
- Root injury or damage leads to infection spreads by root contact and root grafting.
- Fire protection increases disease.
- Heavy growth of weeds favours disease.

Fungus: *Polyporus gilvus*

Management

- Burning reduces soil moisture and check weed growth that reduces diseases incidence.
- All dead, dry stumps and drying trees should be removed.

7. Bacterial wilt

Symptoms

- Small brown patches in between veins that gives a scorched appearance.
- Leads to drooping of leaves-plant become wilted and die.
- Dark discolouration of vascular tissues occurs in roots.
- Infected tissue - immersed in the glass of water, streak of creamy white bacterial ooze.

Bacterium: *Ralstonia solanacearum* (*Pseudomonas solanacearum*)

Management

- Selection of nursery area with good drainage facilities.
- Root injuries should be avoided.
- Soil solarization or fumigation with formalin can be done in nursery area before raising seedlings.

» Spray application of plantomycin 0.01% or copper hydroxide (100 ppm).

» Nursery application of biocontrol agents like *Pseudomonas fluorescens* at 2.5 kg /50 kg FYM.

8. Flowering parasite (Giant mistletoe)

Symptoms

» Produces a very hard flattened prostrating stem that produce haustorial connections to drive the food source.

» Point of parasitism becomes thickened to form a gall like from which multiple shootings (15-20) arise as clusters.

» Affected branch start withering drying from tip and dry off and produce erect, clustered branches from the galled portion.

» The parasite is a shrubby plant with green leaves, produces cluster of orange coloured tubular flowers with dark red coloured berries.

» The birds are attracted by the fruits and act as dissemination agents.

Parasite: *Dendrophthoe falcata, Dendrophthoe lonasa* (Partial stem parasite)

Management

» Collect all the seeds and destroy.

» Remove affected portions at point of attachment and apply Bordeaux paste to cut ends.

9. Flowering parasite (Dodder)

Symptoms

» Dodder identified by its thin stems appearing leafless / leaves reduced to minute scales.

» Dodder flowers are white to pink / yellow to cream.

» After a dodder attaches itself to a plant, it wraps itself around it.

» Produces haustoria that insert themselves into the vascular system of the host.

- The original root of the dodder in the soil then dies.
- The dodder attaches itself to multiple plants.
- May reach high into the canopy of shrubs and trees.

Parasite: *Cuscuta reflexa, Cuscuta campestris* (Total stem parasite)

Management

- Collect all the seeds and destroy.
- Remove affected portions at point of attachment and apply Bordeaux paste to cut ends.

Other Diseases

10. **Seedling blight**: *Pellicularia rolfsii*
11. **Leaf spot**: *Phoma casuarina*
12. **Leaf tip blight**: *Glomerella cingulata*, *Phomopsis* sp.
13. **Black root rot**: *Macrophomina phaseolina*
14. **Root rot**: *Armillaria mellea*
15. **Heart rot**: *Polyporus glomeratus*, *Fomes fastuosus* and *Fomes senex*
16. **Brown rot**: *Phellinus noxius*
17. **Bacterial wilt**: *Ralstonia solanacearum*
18. **Stem canker and die-back**: *Phomopsis casuarinae, Botryosphaeria ribis*

Selected References

Narayanan C, Mark Dudzinski, Sharma JK, Mohanan C. 1996. The Extent, recognition and management of blister bark disease. In: Pinyopusarerk, K., J.W. Turnbull and S.J. Midgley (Eds.) Recent Casuarina research and development. CSIRO Forestry and Forest Products, Canberra, Australia, pp. 74-79.

Narayanan C, Sharma JK, Minter DW. 2003. *Subramanianospora vesiculosa* - a hyphomycete causing wilt disease of *Casuarina equisetifolia*. Indian Phytopathology 56 (2), 159-163.

Ali MIM, Anuratha CS, Sharma JK. 1991. Bacteria wilt of Casuarina equisetifolia in India. Forest Pathology, 21(4), 234–238.

Mohanan C, Sharma JK, 1989. Occurrence of new diseases of *Casuarina equisetifolia* in India. The Indian Forester, 115,1

Mohanan C, Sharma JK, 1993. Diseases of *Casuarina equisetifolia* in India. Commonwealth forest review, 72, 48-52.

Thiribhuvanamala G. 2017. Major diseases affecting pulp wood trees and their management. In: K.T. Parthiban, R. Seenivasan (Eds) Plantation and Agroforestry Pulpwood Value Chain Approach, Scientific Publisher, India

19 Diseases of *Cedrus deodara* (Deodar cedar)

S. No.	Disease	Pathogen
1.	*Seiridium* canker	*Seiridium cardinale, Seiridium cuppressi, Seiridium unicorne*
2.	Root rot	*Phytophthora cinnamomi*

1. Seiridium canker

Symptoms

- » Initially, multiple cankers were formed and girdled branches, severing nutrient transport, and canker wounds may ooze resinous sap.
- » An early symptom of canker disease is tip dieback on individual "flagged branches."
- » The main trunk, which is also called the central leader, starts dying from the top down and leaving brown needles.
- » Although mature trees may take several years to die, seiridium canker disease can kill a young tree in less than a year.

Fungi: *Seiridium cardinale, Seiridium cuppressi, Seiridium unicorne*

Management

- » Remove and destroy the infected plants.

2. Root rot

Symptoms

- The disease has resulted in the decline and death of trees.
- The foliage in the infected trees turns yellowish, the length of needles and shoot is reduced and finally the diseased trees become conspicuous with scant and chlorotic foliage extending in centripetal manner.
- Cone size is also considerably reduced.
- The tops of the trees and tips of the branches dry up.
- Below the diseased patch runs the forest road and below this road all the trees are healthy.

Oomycete: *Phytophthora cinnamomi*

Management

- Remove and destroy the infected plants.
- Spray with Metalaxyl + Mancozeb 0.1%.

Selected References

Tewari D.N. 1994. A Monograph on Deodar (*Cedrus deodara* (Roxb.) G. Don). International Book Distributors, Dehradun, India. 212 pp.

Singh L, Kaur MJ, Tapwali A. 2010. Himalayan Research Group, Umesh Bhavan, Chotta Shimla, Shimla Jorhat Evaluation of chemical and bio-control agents for management of *Cedrus deodara* root rot caused by *Phytophthora cinnamomi*, Indian Phytopath., 63 (1): 59-62.

Singh L, Lakhanpal TN. 2000. *Cedrus deodara* root rot disease-threat to the Himalayan forestry and environment. Indian Phytopath. 53: 50-56.

20 Diseases of *Dalbergia latifolia* (Bombay blackwood)

S. No.	Disease	Pathogen
1.	*Physalospora* leaf spot	*Physalospora dalbergiae*
2.	Tar spot	*Phyllachora dalbergiae*
3.	Leaf rust	*Uredo sissoo*
4.	*Colletotrichum* leaf spot	*Glomerella cingulata*
5.	Powdery mildew	*Phyllactinia dalbergiae*
6.	Seedling die-back	*Colletotrichum capsici*

1. Physalospora leaf spot

Symptoms

» Leaves of all the ages were found to be susceptible to infection.

» The leaf tissue of the affected area became yellowish-green and gradually turned into an amphigenous necrotic spot.

» Black, shiny dot-like ascomata developed in groups on the adaxial surface of these spots.

» In severe cases five to eight such leaf spots could be present on a single leaflet.

Fungus: *Physalospora dalbergiae*

Management

» Spray with Blue copper 0.05% or Mancozeb 0.1% or Captan 0.1%.

2. Tar spot

Symptoms

» The infected leaves showed deep coal tar like raised specks with fainted yellow halo appear on upper surface of leaf where they shiny black, cushiony ascomata developed singly or in clusters and later developed to large sized spots covering entire leaf and rarely appears on lower surface of leaf.

» Leaf affected area turned pale and gradually to light brown colour.

» Mature leaves showed high severity and pre-mature defoliation.

Fungus: *Phyllachora dalbergiae*

Management

» Spray with Carbendazim 0.2% or Propiconazole 0.1% or Fluopyram+Tebuconazole 0.1%.

3. Leaf rust

Symptoms

» Minute, sparsely scattered uredeniosori, rusty brown in colour, appeared on the abaxial surface of leaflets.

» Only in the case of severe infection some light greenish-yellow flecks developed on the adaxial surface.

» Severely rusted leaflets turned yellow and got defoliated prematurely.

Fungi: *Uredo sissoo, Maravalia pterocarpi, Uromyces achorus*

Management

» Spray with Oxycarboxin 0.1%.

4. *Colletotrichum* leaf spot

Symptoms

- Young as well as mature leaves were equally susceptible to infection.
- Small brown amphigenous necrotic spots, 1-2 mm. across, appeared scattered over the leaflets, but commonly along the margin and near the tip.
- Later, these spots coalesced to from large irregular reddish brown spots surrounded by light greenish yellow border.
- The necrotic tissue was often shed leaving shot-holes in the lamina.

Fungus: *Colletotrichum* state of *Glomerella cingulata*

Management

- Spray Mancozeb 0.5 g / lit at weekly intervals.

5. Powdery Mildew

Symptoms

- The fungus produces whitish, persistent, dense mycelium on the lower surface of leaves.
- With the age, the mycelium and conidiophores bearing conidia become grayish-white to pale yellow.
- Most of the leaves on the trees are covered with white colony growth of the fungus, perithecia start to form, which were initially orange in colour.
- The ascocarps turn into brown with the progress of the disease and ultimately become black at maturity.
- The powdery mildew causes severe defoliation but never kills the tree.

Fungus: *Phyllactinia dalbergiae* imperfect stage of *Ovulariopsis sissoo*

Management

- Spray with Wettable sulphur 0.2% or Bitertanol 0.1% or Myobutanil 0.1%.

6. Seedling dieback

Symptoms

- The infected plants started dying from top to downward.
- Plant became partially to completely dry.
- The occurrence of the disease was calculated after counting the numbers of die back disease affected seedlings.

Fungus: *Colletotrichum capsici*

Management

- Spray with Metalaxyl + Mancozeb 0.1%.

Other Diseases

7. **Leaf spot:** *Cercospora dalbergiae-latifoliae*
8. **Root rot**: *Coriolopsis sanguinaria*
9. **Sooty mould**: *Meliola bicornis*
10. **Phanerogamic parasites**: *Dendrophthoe falcata*

Selected References

Bagchee K. 1958. Diseases of forest trees in India—causes, effects and methods of prevention. Indian Forester, 84: 407–417

Bakshi BK. 1976. Forest Pathology - Principles and Practice in Forestry. FRI Press, Dehradun.

Brown A, Green S, Hendry S. 2005. Needle diseases of pine. Forestry Commission Information Note 67. Forestry Commission, Edinburgh.

Jamaluddin Goswami MG, Ojha BM. 2004. Fungi of India (1989-2001). Scientific Publisher, Jodhpur (India) 326p.

Sharma JK, Mohanan C, Florence EJ. 1984. Diseases survey in nurseries and plantations of forest tree species grown in Kerala Forest Research Institute. Peechi, p.36.

Sharma RC, Bhardwaj LN. 1988. Forest nursery diseases and their management. In: Parkash

Ram (ed) Advances in forestry in India, Vol II. International Book Distributors, Dehradun, pp 91–118.

Swamy KR, Annappa, Vinayak Pai, Hanumantha M and Suryanarayan V. 2019. Assessment of tar spot disease in *Dalbergia latifolia* and their management. Journal of Pharmacognosy and Phytochemistry. 8(3): 621-623.

Verma P, Soni KK, Verma KR, Fatima S. 2015. Seedling die back of *Dalbergia latifolia* in central India caused by *Colletotrichum capsici*- A new record. International Journal of Current Microbiology and Applied Sciences, 5(6): 350-356.

Verma RK, Sharma N, Soni KK, Jamaluddin. 2008. Forest fungi of central India. International Book Distributing Co. Lucknow, 418p.

21 Diseases of *Dalbergia sissoo* (Indian rosewood)

S. No.	Disease	Pathogen
1.	Leaf spot	*Cercospora sissoo, Colletogloeum sissoo, Curvularia affinis, Pseudocercospora dalbergiae, Phyllosticta sissoo, Myrothecium roridum, Alternaria alternata*
2.	Tar spot	*Phyllachora dalbergiae, Phyllachora spissa*
3.	Wilt	*Nectria haematococca, Fusarium solani*
4.	Decline	*Botryodiplodia theobromae, Ceratocystis fimbriata, Fusarium solani*
5.	Die-back	*Fusarium solani, Botryodiplodia theobromae, Ganoderma lucidum, Phytophthora cinamomi, Pseudomonas* sp.
6.	*Rhizoctonia* web blight	*Rhizoctonia solani*
7.	Powdery mildew	*Phyllactinia dalbergiae*
8.	*Maravalia* rust	*Maravalia achroa*
9.	*Uredo* rust	*Uredo sissoo*
10.	*Alternaria* leaf blight	*Alternaria alternata*
11.	*Phoma* leaf blight	*Phoma indica*
12.	Anthracnose	*Colletotrichum gloeosporioides*
13.	*Ganoderma* root rot	*Ganoderma lucidum*

1. Leaf spot

Symptoms

Cercospora sissoo

- The pathogen, attacks the leaves mostly on the lower surface, producing yellowish to grayish-green discoloration.
- Pustules are mostly intra-epidermal.
- Stomata are brown with simple or forked conidiophores.

Colletogloeum sissoo

- The pathogen, causes imperceptible leaf spots.

Curvularia affinis

- Leaf spots were mostly irregular, beginning at the leaf margin and moving inwards.
- The colour of leaf spots was light reddish with dark margins and a light yellow halo surrounding the lesion.

Pseudocercospora dalbergiae

- The disease appears as small circular spots on leaves.
- The size of the spots increases with the progress of times.
- Larger spots develop due to coalescence of closure spots mostly on the lower surface of leaves producing yellowish to grayish-green discoloration.
- The mycelium is brown and intra-epidermal.

Phyllosticta sissoo

- The spots are round to irregular, greyish-brown which sometimes cover the entire leaf surface.
- Dark brown pycnidia are produced on lower leaf surface in densely aggregated groups.

Myrothecium roridum

- They are grey or light brown with dark brown margin on the concentric zones, coalescing to form larger leaf spots.

- The necrotic tissues usually fall off resulting in formation of shot holes.

Alternaria alternata

- Infection spots are dark brown, vary in size and shape and coalesce to form larger spots.
- The fungus speculates on the lower surface of the leaves.
- The heavily infected leaves are shed prematurely.

Management

- Spray with Carbendazim 0.1% or Copper oxy chloride 0.05% or Chlorothalonil 0.1% or Mancozeb 0.2%.

2. Tar spot

Symptoms

Phyllachora dalbergiae

- The pathogen, attacks the upper leaf surface and produces shining black cushion-like stromata which may occur scattered or in clusters.

Phyllachora spissa

- The pathogen, attacks the leaves and forms densely aggregated dot-like dark stromata on irregular brownish infection spots.

Management

- Spray with Carbendazim 0.1% or Copper oxy chloride 0.05% or Chlorothalonil 0.1% or Mancozeb 0.2% at fortnightly intervals.

3. Wilt

Symptoms

Nectria haematococca

- Characteristic symptoms of the disease are yellowing and death of leaves in acropetal succession up the tree and affected trees die within a few months.

- It can cause mass mortality, with almost the entire mortality.

Fusarium solani (Fusarium solani* f.sp. *dalbergiae)

- The symptoms are characterized by drooping leaves and branches due to loss of turgor.
- The leaflets gradually turn, yellow, and eventually drop off.
- The entire tree becomes 'thin' and light coloured, and stands out sharply in contrast to adjoining densely clothed green healthy trees.
- The nodules and slender branches on roots which characterize healthy sissoo are absent on the roots of wilted trees.
- On splitting the diseased roots, the bark and the outer sapwood exhibit a well-defined, pinkish-brown, (sorghum brown, army brown, light seal brown) stain, and though the stain is restricted in this region, it may rarely penetrate into the inner sapwood.
- The outer sapwood exhibits a characteristic pink to reddish pink stain.
- The heartwood is free from the stain.
- Though stain also progresses along the outer sapwood from the roots to the stem and, in the late stage of wilt, it may extend up to stem about ten feet from the ground.
- In diseased roots, vessels are plugged with hyphae and infiltration with jelly like substances which hinder sap flow to the crown resulting in wilt symptoms.

Management

- Soil drenching with organic amendments like FYM or vermicompost or coirpith compost along with *Pseudomonas fluorescens* and *Trichoderma viride.*

4. Decline

Symptoms

- Initially the top leaves of infected plants turn light yellow.
- The withering of individual branches from top to downward takes place.
- Bark at the collar region becomes rotted; turn brown and often splits, and roots become dark brown to black.

- Bark of trunk splits, gets separated from cambium and can be peeled off easily.
- There are holes in the bark and galleries in the cambium with dark brown powder like material due to insects that may contribute in the disease spread.
- The whole plant dries up gradually and takes several months to dry completely.

Fungi: *Botryodiplodia theobromae, Ceratocystis fimbriata, Fusarium solani*

Management

- Spray with Carbendazim 0.1% or Thiophanate methyl 0.2%.

5. Die-back

Symptoms

- In infected trees, twigs start dying from tip and downward due to virulent strains of the pathogen which completely destroyed the root system.
- The affected leaves lose their green colour and later turn brown.
- Under severe conditions, the branches start drying one after the other causing death of the entire tree.

Fungi: *Fusarium solani, Botryodiplodia theobromae, Ganoderma lucidum*

Oomycete: *Phytophthora cinamomi*

Fungi: *Pseudomonas* sp.

Management

- Spray with Mancozeb + Metalaxyl 2 g /lit or Fosetyl aluminium 0.2%.

6. *Rhizoctonia* web blight

Symptoms

- The disease first appears on leaves close to the ground as water soaked grayish brown blotches which increase in size with the advancing fungal hyphae and ultimately the entire leaf blade is invaded by the fungus.
- The leaflets show stromatid aggregates on the under surface and eventually turn brown.

- The infected adjoining leaflets often join together by the fungal hyphae as if caught in a spider's web hence the name web blight.
- Leaflets or entire leaf become detached prematurely but they remain clinging to the stem for a considerable period as they are invariably joined together by the fungal hyphae.
- There is a cluster of hyphae at the base of the petiole or petiole.
- The disease spreads laterally through contact of overlapping foliage of the adjoining seedlings resulting in group infection of seedlings in the nursery.

Fungus: *Rhizoctonia solani* an anamorph of *Thanatephorus cucumeris* (Syn: *Pellicularia filamentosa*)

Management

- Provide proper sanitation and weeding.
- Foliar application of Triadimefon 0.1% at fortnightly intervals.

7. Powdery mildew

Symptoms

- The fungus produces whitish, persistent, dense mycelium on the lower surface of leaves.
- With the age, the mycelium and conidiophores bearing conidia become grayish-white to pale yellow.
- Most of the leaves on the trees are covered with white colony growth of the fungus, perithecia start to form, which were initially orange in colour.
- The ascocarps turn into brown with the progress of the disease and ultimately become black at maturity.
- The powdery mildew causes severe defoliation but never kills the tree.

Fungus: *Phyllactinia dalbergiae* imperfect stage of *Ovulariopsis sissoo*

Management

- Spray with Wettable sulphur 0.2% or Bitertanol 0.1% or Myobutanil 0.1%.

8. Maravalia **rust**

Symptoms

- The infection declines following monsoon rains.
- The affected parts are killed resulting in die-back and subsequent death of affected seedlings.
- Uredinial sori are yellowish and formed on the lower surface of the leaves.
- Telia are colourless and pulvinate.
- The infected leaves are often deformed and the infected plants show perceptible retardation in growth and look stunted and weak.

Fungus: *Maravalia achroa*

Management

- Spray with Oxycarboxin 0.2% or Triadimenol 0.1% at fortnightly intervals.

9. Uredo **rust**

Symptoms

- The disease develops as minute yellow uredia on the lower surface of the leaves with chlorotic spots on the upper surface.
- As the disease advances thousands of black raised rough telia also develop on the lower surface.

Fungus: *Uredo sissoo*

Management

- Spray with Oxycarboxin 0.2% or Triadimenol 0.1% at fortnightly intervals.

10. Alternaria leaf blight

Symptoms

- » A brown water soaked lesion develops on the leaf and spreads fast to occupy major portion of the leaf leading to complete drying of the lamina.
- » The disease can easily identified by presence of concentric rings in the blighted area showing the sporulation of the fungus.
- » The fungus produces brown coloured conidia with both vertical and horizontal septa with a beak at the top.

Fungus: *Alternaria alternata*

Management

- » Spray with Carbendazim 0.1% or Thiophanate methyl 0.2%.

11. Phoma leaf blight

Symptoms

- » A black water soaked lesion develops on the leaf and spreads fast to occupy major portion of the leaf leading to complete drying of the lamina.
- » The disease can easily identified by presence of large number of black dots in the infected tissue showing the asexual fruiting bodies, pycnidia.

Fungus: *Phoma indica*

Management

- » Spray with Carbendazim 0.1% or Chlorothalanil 0.1%.

12. Anthracnose/Fruit rot

Symptoms

- » The infection starts as watersoaked lesions, which later turns to minute black spot with dark margins.
- » As the disease progresses, the spots enlarge and cause irregular blighting of the leaf lamina.

» The blighted leaf tissues show a hundreds of minute black coloured fruiting body of the fungus in concentric rings.

» The type of fruiting body is acervulus with a large number of hyaline, oblong, single celled conidia which helps in the secondary spread of the disease.

Fungus: *Colletotrichum gloeosporioides*

Management

» Spray with Carbendazim 0.1% or Chlorothalanil 0.1%.

13. Ganoderma root rot

Symptoms

» Mostly age old trees are highly susceptible.

» The disease prevalent both in heavy and light soils.

» The disease can be visualized as yellowing and dropping of leaves in one or two branches of the tress, subsequently followed by drying of the branches.

» As the disease progresses, the defoliation and drying symptoms are seen on the other branches.

» The fungus grows extensively from the bark and causes white rot of the sap wood which becomes spongy.

» The dried trees show characteristic fruiting bodies of the fungus which arise from the base of the trees at the soil level.

Fungus: *Ganoderma lucidum*

Management

» Spot drenching with Bordeaux mixture 1%.

Other Diseases

14. Angular leaf spot: *Gibberella* sp.

15. Leaf spot: *Chalara* sp., *Cochliobolus lunatus*, *Cylindrocladium scoparium*, *Phomopsis dalbergiae*, *Mycosphaerella dalbergiae*, *Fusarium solani*, *Lasiodiplodia theobromae*, *Memnoniella* sp. and *Tetraploa* sp.

16. **Leaf blight**: *Glomerella cingulata* and *Septothyrella dalbergiae*

17. **Horse hair blight**: *Marasmius equicrinis*

18. **Branch canker**: *Hypoxylan hypomiltum*

19. **Pod rot**: *Catenulaster batistae, Glomerella cingulata, Cladosporium* sp. and *Septothyrella dalbergiae*

20. **Heart rot**: *Phellinus rimosus, Fomes durissimus, Fomes robiniae*

21. **Root rot**: *Oxyporus latemarginatus* (Syn: *Poria ambigua*), *Neocosmospora vasinfecta* var. *africana, Ganoderma applanatum, Phellinus gilvus* and *Coriolopsis sanguinaria*

22. **Wood rot**: *Polyporus ciliatus*

23. **Butt rot**: *Phellinus gilvus*

24. **Root and butt rot**: *Polyporus gilvus*

25. **Mottled sap rot**: *Schizophyllum commune*

26. **Rust**: *Scopellopsis dalbergiae* on *Dalbergia paniculata; Uredo sissoo, Maravalia achroa* and *Maravalia pterocarpi*

27. **Sooty mould**: *Meliola bicornis*

28. **Die-back**: *Lasiodiplodia theobromae, Phytophthora cinnamomi, Fusarium dimerum, Fusarium solani*, and species of *Pestalotia, Botrytrichum, Botrytis, Phoma, Drechslera, Botryodiplodia Alternaria* and *Rhizoctonia*

29. **Seed infection**: *Aspergillus niger, Aspergillus flavus, Aspergillus terreus, Aspergillus* sp., *Alternaria alternata, Chaetomium* sp., *Drechslera australiensis, Fusarium pallidoroseum, Fusarium solani, Fusarium* sp., *Penicillium* sp., *Rhizopus* sp., *Slternaria* sp and *Geotrichum* sp.

30. **Phanerogamic parasites**: *Tolypanthus involucratus* (Syn: *Loranthus involucratus*), *Loranthus pulverulentus, Dendrophthoe falcata, Cuscuta reflexa, Tapinanthus dodoneifolius.*

Selected References

Bagchee K. 1958. Diseases of forest trees in India—causes, effects and methods of prevention. Indian Forester, 84: 407–417

Bajwa R, Mukhtar I. 2006. Incidence of shisham dieback disease in different agro-ecological zones of the Punjab [Pakistan]. In: Proceedings of third national seminar on Shisham dieback, 11 May 2006, pp 89–106.

Bakhshi BK. 1957. Wilt disease of shisham (*Dalbergia sissoo* Roxb.). The effect of soil moisture on the growth and survival of *Fusarium solani* in the laboratory. Indian Forester, 83: 505-511.

Bakshi BK. 1954. Wilt disease of shisham (*Dalbergia sissoo*) due to *Fusarium solani* Sensu Synder and Hansen. Nature, 174:278

Bakshi BK. 1963. Dangerous forest diseases in India. In: Internationally dangerous forest diseases. Miscellaneous publication 939. Forest service, United States Department of Agriculture, pp 55–61

Bakshi BK. 1976. Forest Pathology - Principles and Practice in Forestry. FRI Press, Dehradun.

Dahiya SS, Sharma JC, Kaushik JC. 2004. Dieback of Shisham (*Dalbergia sissoo* Roxb.) and Kikar (*Acacia nilotica* L.) trees in Haryana (India). Ann Agric Bio Res 9:175–182.

Khan SM, Shakir AS, Tabssum MA, Rehman A. 2001. Isolation and identification of different fungi from diseased shisham tree. Proc. 3rd Natl. Conf. of Pl. Pathol. Oct. 1-3, 2001. NARC, Islamabad. pp. 44-46.

Manadhar G, Shrestha SK, Appanah S, Allard G, Amatya SM. 2000. Fungi associated with dieback of Sissoo. Proc. Intl. Sem., Nepal (18): 27- 29.

Mukerji KG, Bhasin J. 1986. Plant diseases in India. A source Book. Tatta Mc.Grew-Hill Publishig Company Ltd. New Delhi. 467 pp.

Parajuli AV, Bhatta BM, Adhikary K, Tuladhar J, Thapa B. 1999. Causal agents responsible for the dieback of *Dalbergia s*issoo Roxb., in the eastern Nepal Terai. Ban Ko Jankari. 9: 7-14.

Pathan MA, Rajput NA, Jiskani MM, Wagan KH. 2007. Studies on intensity of shisham dieback in Sindh and impact of seed-borne fungi on seed germination. Pak J Agric Agric Eng Vet Sci 23:12–17

Poussio GB, Kazm MR, Akem C, Fateh FS. 2010. First record of *Ceratocystis fimbriata* associated with shisham (*Dalbergia sissoo*) decline in Pakistan. Australas Plant Dis Notes 5:63–65

Rajput NA, Pathan MA, Jiskani MM, Rajput AQ, Arain RR. 2008. Pathogenicity and host range of *Fusarium solani* (Mart.) Sacc., causing dieback of Sisham (*Dalbergia sissoo* Roxb.). Pak. J. Bot., 40(6): 2631-2639.

Shakya DD, Lakhey PB. 2007. Confirmation of *Fusarium solani* as the causal agent of die-back of *Dalbergia sissoo* in Nepal. Plant Pathol., 56:1041 (Oxford)

Shamsi S, Razia Sultana Rumana Azad. 2008. New record of *Phyllactenia dalbergae* Piroz. and its anamorph *Ovulariopsis sissoo* sp. nov. on *Dalbergia sissoo* Roxb. from Bangladesh. Bangladeh J. of Plant Pathol. 24 (1and2): 87-89.

Sharma RC, Bhardwaj LN. 1988. Forest nursery diseases and their management. In: Parkash Ram (ed) Advances in forestry in India, Vol II. International Book Distributors, Dehradun, pp 91–118.

Sutton BC. 1980. The Coelomycetes, Fungi Imperfect with pycnidia Acervuli and Stroma. Commonwealth Mycological Institute. Kew Surrey, England. pp. 525-537.

22 Diseases of *Dendrocalamus strictus* (Bamboo)

S. No.	Disease	Pathogen
1.	Damping-off	*Fusarium moniliforme*
2.	Web blight	*Thanatephorus cucumeris*
3.	Leaf blight	*Helminthosporium turcicum*
4.	Leaf rust	*Dusturella divina*
5.	Basal culm rot	*Fusarium moniliforme* var. *intermedium*
6.	Growing culm rot	*Fusarium equiseti*
7.	Seedling rhizome rot	*Rhizostilbella hibisci*
8.	Culm blight	*Sarocladium oryzae*
9.	Branch die-back	*Fusarium pallidoroseum*
10.	Witches broom	*Balansia liniaris*
11.	Thread blight	*Botryobasidium salmonicolor*
12.	Foliage blight	*Bipolaris maydis*
13.	Leaf tip blight	*Alternaria alternata*
14.	*Exserohilum* leaf spot	*Exserohilum rostratum*
15.	Zonate leaf spot	*Dactylaria* sp.
16.	*Colletotrichum* leaf spot	*Colletotrichum gloeosporioides*
17.	*Phoma* leaf spot	*Phoma sorghina*
18.	*Phomopsis* leaf spot	*Phomopsis* sp.

19.	*Stagonospora* leaf spot	*Stagonospora* sp.
20.	*Chaetospermum* leaf spot	*Chaetospermum carneum*
21.	*Curvularia* leaf spot	*Curvularia lunata*
22.	*Ascochyta* leaf spot	*Ascochyta* sp.
23.	*Petrakomyces* leaf spot	*Petrakomyces indicus*
24.	*Rosenscheldiella* leaf spot	*Rosenscheldiella* sp.
25.	*Cerodothis* leaf spot	*Cerodothis aurea*
26.	*Coccodiella* leaf spot	*Coccodiella* sp.
27.	Tar spot	*Phyllachora ischaemi*
28.	Culm Staining and Die-Back	*Apiospora bambusae*
29.	Sooty mould	*Capnodium braziliense*
30.	Little leaf	*Candidatus* Phytoplasma sp.
31.	Mosaic	*Bamboo mosaic virus*

1. Damping-off

Symptoms

» The diseases is common in seedbed nurseries which affects the emerging seedlings during germination (pre-emergence damping-off) or after germination (postemergence damping-off), while the seedling tissues are still succulent.

» The disease occurs in patches in the seedbeds 7 to 12 days after sowing.

» The disease is characterized by the rotting of seeds and also the radicle.

» Post-emergence damping-off is characterized by development of water-soaked brown lesions on the emerging plumule near the soil level and collapse of the affected plumule in due course.

Fungi: *Fusarium moniliforme*, *F. oxysporum*, *Rhizoctonia solani*

Management

» The disease can be managed by adopting proper nursery cultural practices.

» Low sowing rate, i.e., 500 g seeds (*B. bambos*) per standard seedbed (12 x 1 x 0.3 m) is preferable to prevent the build-up of conditions conducive to the spread of the disease.

- » Excessive watering and shade over the nursery beds should be avoided.
- » Seed dressing with fungicides such as Thiram or Captan at 2 g/kg of seeds
- » Spray fungicide like Carboxin at 0.1% in the affected nursery beds.
- » Soil drenching with copper oxy chloride or with carbendazim 0.1%.

2. Web blight

Symptoms

- » Infection occurs as water-soaked lesions on seedlings stem near the soil level and later spreads rapidly affecting the entire shoot, except one or two juvenile leaves.
- » The infected seedling stem and foliage become discoloured and necrotic.
- » Under high humidity, mycelia of the causal fungus (*R. solani*), arise from the soil and grow epiphytically over the affected seedlings.
- » Yellowish brown sclerotia and basidial stage of the fungus also develop on the decayed basal foliage and stem.
- » The disease occurs in patches and the affected seedlings are killed outright within 10-20 days of infection, leaving large circular to irregular patches of dried-up seedlings in the seedbed.

Fungus: *Thanatephorus cucumeris* basidial stage of *Rhizoctonia solani*

Management

- » Fumigation / soil solarisation of nursery bed.
- » The disease can be controlled lowering the sowing rate (500 g seeds per standard seed beds) and also by lowering the watering regime (120 l per standard beds) in the nursery.
- » Spray with Carboxin 0.2% after seven and 21 days of seedling emergence is recommended for managing the disease.
- » Foliar spraying of fungicides like, Triademefon 0.1% or Carbendazim 0.1% at fortnightly intervals.

3. Leaf blight

Symptoms

- » Appears in August and in November disease becomes severe.
- » Minute spindle shaped water soaked lesions appear on the margin of the leaf blade.
- » Later extends and occupies major portion of the leaf, giving a blighted appearance.

Fungus: *Helminthosporium turcicum*

Management

- » Removal of affected leaf portions.
- » Foliar spraying of Carbendazim 0.1% or Copper oxy chloride 0.05% or Chlorothalonil 0.2% or Mancozeb 0.2%.

4. Leaf rust

Symptoms

- » The rust infection appears as minute pin-head, water-soaked flecks on the adaxial surface of the foliage, where yellowish orange to rust brown linearly arrangedurediniosori develop.
- » On the corresponding abaxial surface of the flecks, grayish brown to dark brown lesions with yellowish orange halo forms.
- » Often numerous such lesions develop on a single lamina which later coalesce and spread to entire leaf lamina.
- » Severe infection leads to necrosis of the leaf tissues between the spots.
- » Rust uredinia develop and continue to produce bright yellowish orange coloured urediniospores.
- » Rust teliosori develop linearly on the adaxial surface of the leaf either in the degenerating urediniosori or separately.
- » Severe rust uredinial infection causes abnormal leaf fall even before the development of the teliosori.

Fungus: *Dusturella divina, Puccinia bambusicola*

Management

» The disease can also be controlled by avoiding shade over the nursery beds, lowering the sowing rate (500 g seeds per standard seed beds) and also by lowering the watering regime (120 l per standard beds) in the nursery.

» Application of fungicide like Oxycarboxin 0.1% or dusting with sulphur-based fungicides can control the disease.

» Spray with Carboxin 0.2% after seven and 21 days of seedling emergence is recommended for managing the disease.

» Foliar application of Mancozeb 0.1% or Triademenol 0.1% or Triforine 0.1%.

5. Basal culm rot

Symptoms

» The disease manifests as dark brown lesions on the outermost culm sheath of the emerging culm (15-20 cm height), near the soil level.

» These lesions spread rapidly and cover the entire outer culm sheaths.

» The infection causes rot of the tender, succulent emerging shoot which becomes discoloured and gives off a strong smell of molasses.

» The disease affects further development of the culm and causes total decay.

Fungus: *Fusarium moniliforme* var. *intermedium*

Management

» Removal debris around the clumps before the onset of monsoon, light burning of the debris over the ground.

» Loosening the soil around the clump before the culm emergence.

» Pruning and removal of branches from the basal part of the older culms during the dry period.

» To avoid mechanical damage to the emerging culms caused by cattle and other animals, clump tending and cleaning operations are recommended only in well-protected stands.

- Application of *Pseudomons fluorescens* or *Trichoderma viride* at 200 g/ plant (2.5 kg mixed with FYM).
- Application of Carbendazim 0.1% or Mancozeb 3 g/ lit is also recommended.
- Soil drenching with Carbendazim 0.1% at monthly intervals

6. Growing culm rot

Symptoms

- The disease appears as water-soaked brown lesions at the base of culm sheaths where they are attached to the nodes.
- Injury on culm sheaths and culms at nodal region, made by the sap sucking insect *Purohita cervina* distant, predisposes the culm to fungal infection.
- Sap oozes out from the pinprick wounds made by the insect, and infection develops in and around these wounds and forms large necrotic lesions.
- The infection often spreads to the entire culm sheath and to the tissues beneath the culm sheath.
- Severely affected culms cease to grow, become shrivelled, and fall off even before they complete their elongation phase.
- Twisting and bending of culms due to severe necrosis on one side of the culm, partial development of branches, breaking of culms at the point of infection, etc.
- The *Fusarium* sporulates profusely on the necrotic tissues of the culm internode and culm sheath.
- Build-up of insect (*P. cervina*) population at the culm elongation phase was found responsible for the spread of disease within the individual culm or among the culms and clumps by way of dispersal of fungal spores mechanically.

Fungus: *Fusarium equiseti* and *Fusarium moniliforme*

Management

- Spray insecticide, Monocrotophos 0.01% against the insect (*Purohita cervina*) in the bamboo stands during the culm elongation phase.

- Spray fungicides Carbendazim 0.2% or Mancozeb 0.2% on the infected culms.

7. Seedling rhizome rot

Symptoms

- General wilting of seedlings, rolling up of young leaves, yellowing, premature defoliation.
- Affected seedling shows dark yellowish brown discolouration-decay of the growing portion of the rhizome buds.
- Infected rhizome buds discoloured and decayed, infection spreads to the entire rhizome of the seedling diseased seedling is killed outright.
- Soil borne- enters the rhizome tissues through injuries caused in roots and rhizomes.
- The fungus proliferates on the roots and the fleshy rhizome buds as well as on other tissues.
- Entry through mechanical wounds and water-logging condition aggravates disease.

Fungus: *Rhizostilbella hibisci*

Management

- Proper hygienic maintenance of nursery.
- Soil drenching with Copper oxychloride 0.2%.

8. Culm blight

Symptoms

- It occurs in well-established older clumps, aged more than 8-10 years.
- Culms which survive the first growing season remain healthy and the spread of the disease between clumps of bamboos is slow.
- The disease results in a sequential die-back of culms in their first season of growth.
- Symptoms appear when culms are nearing full growth.
- The initial symptoms of blight are premature death of culms sheaths and partial collapse of the fragile apical regions.

- » Later, wet rotten patches develop on the internodes, often associated with insect damage.
- » The necrotic patches spread rapidly in the succulent internodes and eventually become confluent.
- » At the same time, symptoms begin to develop in the lower, more fibrous internodes, and spread slowly downwards, resulting in die-back.
- » Epicormic branches from the nodes of the infected culms develop occasionally which also become affected by the disease in due course.

Fungus: *Sarocladium oryzae*

Associated fungi: *Coniothyrium fuckelii*, *Fusarium* spp., *Acremonium strictum, Pteroconium* sp. and *Arthrinium* sp.

Management

- » Weeds and bushes around the clumps should be removed as these act as retainers of moisture which helps in initiation of the infection.
- » Adopt optimum spacing.
- » Soil or debris from the infected clumps or nearby areas should not be transferred to healthy clumps.
- » Drenching the soil around the bamboo clumps with Copper oxychloride and Mancozeb.
- » Soil application of Gypsum in 2 equal splits (500 kg/ha) reduce the sheath rot incidence.
- » Foliar spray of Carbendazim 0.1% or Mancozeb 0.1% or Chlorothalonil 0.1% at boot leaf stage and 15 days later.
- » Spray with Carbendazim combined with Mancozeb or with Blue copper at initial infection.

9. Branch die-back

Symptoms

- » Culms in young developing clumps were found severely affected by the disease.

- The disease occurred during rainy season and became severe during pot-rainy months causing die-back of branches.
- The infection occurs on branches and on three to five internodes at top of young culm in the form of small greyish magenta coloured linear lesions which later develop into necrotic streaks.
- Infection occurs on foliage as pale yellowish lesions, later spreading to the entire lamina, resulting in leaf necrosis, withering and subsequent premature defoliation.
- Under high humidity fungus sporulates and spreads from the branches to the culm node and internodes downwards.

Fungus: *Fusarium pallidoroseum*

Management

- Soil or debris from the infected clumps or nearby areas should not be transferred to healthy clumps.
- Spray fungicides Carbendazim0.2% or Mancozeb 0.2% on the infected culms.

10. Witches broom

Symptoms

- The disease manifests as development of numerous highly shortened shoots at the nodes of mature culms.
- These abnormal shoots develop into highly reduced shoots successively from their nodes.
- The culm sheaths which cover the internodes also become shortened in size and become boat shaped, often with a prominent ligule.
- Successive development of a large number of thin wiry shoots in tuft from the nodes of the infected culms gives rise to the characteristic appearance of witches' broom.
- New shoots emerging from the infected rhizome also show pronounced brooming symptoms.
- Shining black fructifications of the causal fungus develop on the affected shoots after 5-6 months of infection.

Fungus: *Balansia liniaris*

Management

- Removal and burning the infected culms and witches' brooms.
- As the disease is systemic, rhizome or culms from diseased clumps should not be used for vegetative propagation.

11. Thread blight

Symptoms

- Large water-soaked greyish lesions occur on leaves which advance towards the leaf tip.
- Fine silvery white mycelial strands of the causal fungus appear on the lower surface corresponding to the lesions on the foliage.
- Spread of the disease is mainly through physical contact of the advancing fungal hyphae on the diseased foliage with healthy neighbouring foliage.
- Diseased foliage stuck closely together by the mycelial weft of the fungus at the leaf margins, tips and bases.
- Infection causes browning and necrosis, leading to blight of the culm branches, especially foliage.

Fungus: *Botryobasidium salmonicolor*

Management

- Pruning the diseased branches from the affected clumps and cleaning and burning the debris from the ground around the clumps.

12. Foliage blight

Symptoms

- Infection occurs as small water-soaked grayish brown spindle-shaped lesions with pale yellow halo on both young and mature leaves.
- The lesions coalesce and form large spreading grayish brown to yellowish brown irregular lesions with dark brown border, often covering the entire leaf lamina.

» Severe infection causes withering of leaves and defoliation.

Fungus: *Bipolaris maydis* anamorph of *Cochliobolus heterostrophus*

Management

» Spray with Carbendazim + Mancozeb 0.2% on the infected culms.

13. Leaf tip blight

Symptoms

» The infection appears as greyish brown linear to spindle-shaped lesions near margin, base and tip of both young and mature leaves.

» Usually, the leaf tips were found severely affected which led to leaf tip blight.

Fungus: *Alternaria alternata*

Management

» Foliar spraying of Carbendazim 0.2% or Copper oxy chloride 0.2% or Chlorothalonil 0.2% or Mancozeb 0.2%.

14. Exserohilum leaf spot

Symptoms

» The leaf spots occur on mature leaves.

» Infection manifests as small water-soaked greyish black linear to irregular lesions which later coalesced and spread to the entire leaf lamina; infection causes necrosis, withering and premature defoliation.

Fungi: *Exserohilum rostratum* anamorph of *Setosphaeria rostrata* and *Exserohilum holmii* anamorph of *Setosphaeria holmii*

Management

» Foliar spraying of Carbendazim 0.2% or Copper oxychloride 0.05% or Chlorothalonil 0.1% or Mancozeb 0.2%.

15. Zonate leaf spot

Symptoms

- Infection occurs usually on foliage in the lower branches as minute greyish brown spot of 2-3 mm across.
- The spots enlarge to 5-8 mm in dia and become yellowish brown with dark brown margin.
- The spots spreads further and formed reniform, semi-circular to circular greyish brown areas of 5-10 mm in width with dark brown wavy margin around the pale color central spot; later, these develop into a large zonate spot of 3-12 mm in dia.

Fungus: *Dactylaria* sp.

Management

- Foliar spraying of Carbendazim 0.1% or Copper oxy chloride 0.05% or Chlorothalonil 0.2% or Mancozeb 0.2%.

16. Colletotrichum leaf spot

Symptoms

- The disease manifests as small water-soaked grayish brown lesions on the juvenile and mature leaves.
- These spots spread and coalesce to form large purple linear to irregular areas which often cover the entire leaf lamina, as in *O. travancorica* or *O. ebracteata.*
- Infected leaves become pale yellowish and leathery.

Fungus: *Colletotrichum gloeosporioides* anamorph of *Glomerella cingulata*

Management

- Foliar spraying of Carbendazim 0.1% or Chlorothalonil 0.1% or Mancozeb 0.2%.

17. Phoma leaf spot

Symptoms

- Infection occurs as small pin-head sized brown lesions on the adaxial surface of the leaves. Both the juvenile and mature leaves were found affected with the disease.

- The spots become spindle-shaped and later coalesce to form large irregular spot with greyish white centre and dark brown margin.
- The spots develop and the pycnidia of the fungus formed in the necrotic lesions as erumpent structures.
- Under high humidity, pale pink coloured gelatinous fungal spore mass produced in cirii from the pycnidia.

Fungi: *Phoma sorghina, Phoma herbarum*

Management

- Foliar spraying of Carbendazim 0.2% or Chlorothalonil 0.1% or Mancozeb 0.1%.

18. Phomopsis leaf spot

Symptoms

- The infection manifests as minute greyish brown water-soaked lesions on the mature leaves which later spread to form circular to irregular spots with dark brown wavy margin.
- Pycnidia developed in the necrotic tissues and conidial mass oozed out in yellowish color.

Fungus: *Phomopsis* sp.

Management

- Foliar spraying of Carbendazim 0.1% or Chlorothalonil 0.1% or Mancozeb 0.1%.

19. Stagonospora leaf spot

Symptoms

- The disease affects matured leaves.
- The infection appears as dark brown irregular lesions of 3-5 mm diameter which later enlarge and become brownish black necrotic spots.
- The spots usually develop along the leaf margins.

Fungus: *Stagonospora* sp.

Management

» Foliar spraying of Copper oxychloride 0.05% or Mancozeb 0.1%.

***20. Chaetospermum* leaf spot**

Symptoms

» The disease manifests as numerous minute pale yellow lesions arranged linearly on the upper surface of the mature leaves.

» High humidity and presence of water drops on the leafsurface possibly help in rapid spread of the lesions.

» Usually, development of large number of such lesions on the leaf imparted yellowish colour to the affected foliage.

» Yellowish brown minute pycnidia developed in the necrotic areas during October-November.

Fungus: *Chaetospermum carneum*

Management

» Foliar spraying of Carbendazim 0.1% or Mancozeb 0.1%.

***21. Curvularia* leaf spot**

Symptoms

» The disease affects only the juvenile foliage of new culms.

» The leaf spots appear as greyish black irregular lesions on the juvenile expanding foliage, especially those in the lower branches of new culm.

» The lesions enlarge and cover the entire leaf lamina and become necrotic.

Fungi: *Curvularia lunata* anamorph of *Cochliobolus lunatus*; and *Curvularia pallescens.*

Management

» Foliar spraying of Chlorothalonil 0.1% or Mancozeb 0.1%.

22. Ascochyta leaf spot

Symptoms

- Infection occurs as minute spindle-shaped yellowish brown to brown water soaked spots on the upper surface of the leaves.
- Both juvenile and mature leaves were found affected by the disease.
- Dark brown to black pycnidia develop over the necrotic area from which under high humidity pinkish white spore mass oozed out.
- Severe infection caused premature defoliation.

Fungus: *Ascochyta* sp.

Management

- Foliar spraying of Carbendazim 0.1% or Chlorothalonil 0.1% or Mancozeb 0.1%.

23. Petrakomyces leaf spot

Symptoms

- The disease manifests as pin-head sized brown water-soaked lesions on the foliage, especially those on the lower branches of the culms.
- The lesions enlarged to form 3-5 mm dia. oval to elliptical dark violet coloured spot with pale yellow halo.
- Later, the spots appeared as raised black structures bearing pycnidia of the causal fungus.

Fungus: *Petrakomyces indicus*

Management

- Foliar spraying of Copper oxy chloride 0.05% or Chlorothalonil 0.1% or Mancozeb 0.1%.

24. Rosenscheldiella leaf spot

Symptoms

- The infection appears as minute yellowish brown linear lesions on the mature leaves which in due course enlarged to form 3-5 mm dia. necrotic spots with yellow halo.

» The fungal fructifications produced in linear rows over the necrotic area on the upper surface of the leaves.

Fungus: *Rosenscheldiella* sp.

Management

» Foliar spraying of Carbendazim 0.1% or Chlorothalonil 0.1% or Carbendazim + Mancozeb 0.1%.

25. Cerodothis leaf spot

Symptoms

» The infection affects both juvenile and mature leaves and manifests as tiny, pale yellow spots on the upper surface of the leaves.

» No visible necrotic areas were formed as the disease progressed.

» The ascocarps of the fungus become erumpent through the ruptured epidermis.

» At maturity, the ascocarps appeared astiny golden yellow streaks arranged in linear rows.

» Under warm-humid condition, the entire leaf surface becomes covered with the spore masses.

Fungus: *Cerodothis aurea*

Management

» Foliar spraying of Chlorothalonil 0.1% or Carbendazim + Mancozeb 0.1%.

26. Coccodiella leaf spot

Symptoms

» The infection appears as yellowish brown minute lesions which enlarged to form dark brown linear necrotic spot.

» Fructifications of the causal fungus developed in the necrotic spot on the lower surface of the leaves.

Fungus: *Coccodiella* sp.

Management

- Foliar spraying of Carbendazim 0.1% or Azoxystrobin 0.1% or Chlorothalonil 0.1% or Carbendazim + Mancozeb 0.1%.

27. Tar spot

Symptoms

- Infection appeared as pin-head sized pale to dark yellowish brown lesions on the abaxial surface of the leaf.
- The lesions spread and developed into oval to circular spots with dark brown centre and pale yellow margin.
- Usually, four to six small spots (3-6 mm dia.) appeared on the leaf lamina, as well as on leaf sheath.
- Ascocarps developed as dark brown to black raised structures in the necrotic spots.

Fungi: *Phyllachora ischaemi*, *P. longinaviculata* and *P. shiraiana*.

Management

- Foliar spraying of Tebuconazole 0.1% or Copper oxy chloride 0.05% or Azoxystrobin 0.1% or Carbendazim + Mancozeb 0.1%.

28. Culm Staining and Die-Back

Symptoms

- Infection appeared as pin-head sized pale to dark yellowish brown lesions on the abaxial surface of the leaf.
- The lesions spread and developed into oval to circular spots with dark brown centre and pale yellow margin.
- Usually, four to six small spots (3-6 mm dia.) appeared on the leaf lamina, as well as on leaf sheath.
- Ascocarps developed as dark brown to black raised structures in the necrotic spots.

Fungus: *Apiospora bambusae*

Management

» Foliar spraying of Carbendazim 0.1% or Azoxystrobin 0.1% or Chlorothalonil 0.1% or Carbendazim + Mancozeb 0.1%.

29. Sooty mould

Symptoms

» The infection appears on the upper surface of the leaf as a sparse, black network of hyphae.

» Then a thin effuse black powdery coating can be seen covering a major portion of the leaf. This indicates the sporulation of the fungus.

» The heavy infestation of the sucking pests lead to development of the sooty growth on the leaf surface that blocks sunlight affects photosynthesis and results in stunted plant growth.

Fungus: *Capnodium braziliense*

Management

» Cutting and burning the severely infected leaves.

» Spray starch solution 5% on the infected foliage.

30. Little leaf

Symptoms

» The disease characterized by the development of numerous, highly reduced, abnormal bushy shoots from the nodes of newly emerged culms and culm branches.

» Foliage develops from these shoots show prominent reduction in size and needle-like appearance.

» Profuse development of such abnormal shoots from each node of the developing culm and their subsequent growth gives rise to a massive bushy structure around each node.

» The disease also affects culm elongation; infected culm shows stunted growth.

Bacterium: *Candidatus* Phytoplasma sp.

Management

- » Cutting and burning the severely infected culms/clumps and planting disease resistant bamboo species.
- » As the disease is systemic, rhizome or culms from diseased clumps should not be used for vegetative propagation.

31. Mosaic (Seedling Leaf Striping and Stunting)

Symptoms

- » Pale yellowish to greenish white stripes occur on both young and mature leaves.
- » Often the individual stripes merge together and the leaves become greenish white and leathery.
- » Affected seedlings show stunted growth, and their stem becomes thin, fragile, pendulous and easily breakable.
- » New shoots developed from the rhizome also show similar disease symptoms.
- » Usually, the viruses that cause leaf striping, seedling stunting and mosaic diseases are transmitted mechanically, through seeds or vegetative propagules.
- » In bamboo nurseries, disease may be transmitted through seeds; planting stock from the diseased nurseries should be subjected to thorough screening and seedlings exhibiting mild disease symptoms should be discarded.

Virus: *Bamboo mosaic virus* (BMoV) *Potex virus*

Management

- » Strict quarantine measures against the movement of infected planting materials from the diseased areas should be followed.

Other Diseases

32. **Infection of inflorescence and seed**: *Bipolaris* sp., *Exserohilum* sp., *Fusarium pallidoroseum*, *Drechslera* sp., *Phomopsis* sp., etc.

33. Seedling leaf blight and leaf spots: *Exserohilum rostratm*, *E. holmii*, *Bipolaris maydis*, *B. urochloae*, *Bipolaris* sp., *Dactylaria* sp., *Alternaria alternata*, *Curvularia pallescens* anamorph state of *Cochliobolus pallescens*, and *Colletotrichum gloeosporioides*.

Selected References

Azmy HM, Maziah Z. 1990. Leaf diseases of bamboo. FRIM Technical Information No. 18. Kuala Lumpur, Forest Research Institute of Malaysia.

Bakshi BK. 1976. Forest Pathology, Principles and Practices in Forestry. Controller of Publication, New Delhi, India, p.400.

Bhat MN, Hegde RK, Hiremath PC, Kulkarni S. 1989. Unrecorded pathogen on bamboo causing blight in India. Current Science, 58(2): 1148-1149.

Boa ER, Brady BL. 1987. Sarocladium oryzae associated with a blight of *Bambusa* spp. in Bangladesh. Transactions of the British Mycological Society, 89(2): 161-166.

Chen CC. 1971. Some studies on witches' broom of trees and bamboos in Taiwan. Memoirs of the College of Agriculture (National Taiwan University), 12(2):67-82.

Chen TH. 1985. Bamboo mosaic virus associated with a mosaic disease of bamboos in Taiwan. Plant Protection Bulletin, 27(2):111-116.

Deka PC, Baruah G, Devi M. 1990. A preliminary investigation of diseases of bamboo in North-East region of India. Indian Forester, 116(9): 714-716.

Lin NS, Chai YL, Huang TY. 1993. Incidence of bamboo mosaic potex virus in Taiwan. Plant Disease, 77(5): 448-450.

Mohanan C. 1990. Diseases of bamboos in Kerala. In: International Bamboo Workshop, Cochin 14-18 November 1988. Bamboos - Current research: Proceedings edited by I.V.R. Rao; R. Gnanaharan and C.B. Sastry. Peechi, Kerala Forest Research Institute. pp. 173-183.

Mohanan C. 1993a. Epidemiology and control of *Rhizoctonia* web blight of bamboo. In: IUFRO Symposium, Peechi, 23-26 November 1993. Impact of diseases and insect pests in tropical forests: Proceedings edited by K.S.S. Nair and J.K. Sharma. Peechi, KFRI. pp. 169-185.

Mohanan C. 1994a. Studies on diseases of bamboos and nursery management of *Rhizoctonia* web blight in Kerala. Ph.D. thesis. Cochin University of Science and Technology, Cochin.

Mohanan C. 1994b. Diseases of bamboos and rattans in Kerala. KFRI Research Report No.98. Peechi, KFRI. 120p.

Mohanan C. 1994c. Little leaf disease of bamboos. Bamboo Information Centre India Bulletin, 4(1-2): 30-37.

Mohanan C. 2000. Epidemiology and integrated management of web blight in the bamboo nurseries in India. In: 4 Meeting of IUFRO Working Party 7.03.04 Diseases and Insects in Forest Nurseries. Proceedings. Finland, Finnish Forest Research Institute. pp. 107-120.

Mohanan C. 2004. Witches'broom disease of reed bamboo in Kerala, India. Forest Pathology, 34(5): 329-333.

Mohanan C. 2010. Rust fungi of Kerala. KFRI Handbook No. 26. Peechi, KFRI. 148p.

Mohanan C., Liese, W. 1990. Diseases of bamboo. International Journal of Tropical Plant Diseases, 8 (1): 1-20.

Mohanan, C. 1997. Diseases of bamboos in Asia: An illustrated manual. New Delhi, INBAR. 228p.

Sharma JK, Mohanan C, Florence EJ. 1984. Diseases survey in nurseries and plantations of forest tree species grown in Kerala Forest Research Institute. Peechi, p.36.

Streets RB. 1982. The Diagnosis of Plant Diseases: a field and laboratory manual emphasizing the most practical methods for rapid identification. The University of Arizona Press: Tucson, Arizona.

Thiribhuvanamala G. 2017. Major diseases affecting pulp wood trees and their management. In: K.T. Parthiban, R. Seenivasan (Eds) Plantation and Agroforestry Pulpwood Value Chain Approach, Scientific Publisher, India

Zhu XQ. 1989. The study on witches' broom of bamboos (III): Infection characteristics and control experiment of germs. Journal of Nanjing Forestry University, 13(2): 46-51.

23 Diseases of *Eucalyptus* spp. (Blue gum)

S. No.	Disease	Pathogen
1.	Damping-off	*Pythium* spp.
2.	Vascular wilt	*Fusarium oxysporum*
3.	*Cylindrocarpon* root rot	*Cylindrocarpon lucidum*
4.	Woody stem and root rots	*Phellinus noxius*
5.	Leaf web blight	*Rhizoctonia solani*
6.	Rust	*Austropuccinia psidii*
7.	Powdery mildew	*Sphaerotheca pannosa*
8.	Black mildew	*Meliola densa*
9.	*Cryptosporiopsis* leaf and shoot blight	*Cryptosporiopsis eucalypti*
10.	*Cylindrocladium* foliar spot and foliar blight	*Calonectria hederae*
11.	*Cylindrocladium* shoot blight	*Calonectria quinueseptata*
12.	*Chrysoporthe* stem canker	*Chrysoporthe cubensis*
13.	*Amphilogia* stem canker	*Amphilogia gyrosa*
14.	*Coniothyrium* stem canker	*Coniothyrium zuluense*
15.	*Cytospora* stem canker	*Cytospora eucalypticola*
16.	*Valsa* stem canker	*Valsa eucalypti*
17.	*Macrovalsaria* stem canker	*Macrovalsaria megalospora*
18.	*Thyronectria* stem canker	*Thyronectria pseudotricha*
19.	*Hysterium* stem canker	*Hysterium angustatum*
20.	*Nattrassa* stem canker	*Nattrassa toruloidea*

21.	Pink disease	*Phanerochaete salmonicolor*
22.	*Lasiodiplodia* stem (root collar) canker	*Lasiodiplodia theobromae*
23.	*Seiridium* stem canker	*Seiridium eucalypti*
24.	*Mycosphaerella* leaf spot	*Mycosphaerella cryptica*
25.	*Aulographina* leaf spot	*Aulographina eucalypti*
26.	*Pseudocercospora* leaf spot	*Pseudocercospora acerosa*
27.	*Sonderhenia* leaf spot	*Sonderhenia eucalypticola*
28.	*Septoria* leaf blight	*Phaeophleospora eucalypti*
29.	*Phaeoseptoria* leaf spot	*Phaeoseptoria eucalypti*
30.	*Coniella* leaf blight	*Coniella australiensis*
31.	*Guignardia* leaf spot	*Guignardia citricarpa*
32.	*Trimmatostroma* leaf spot	*Trimmatostroma bifarium*
33.	Bacterial wilt	*Ralstonia solanacearum*
34.	Bacterial leaf blight	*Xanthomonas campestris* pv. *eucalypti*
35.	Little leaf blight	*Candidatus* Phytoplasma
36.	Mosaic	Virus

1. Damping-off

Symptoms

- » Soil borne water mould systemic disease in nursery, causes pre emergence damping off where seeds rot before emergence and post emergence damping off where seedlings rot after they emerge.
- » Pre-emergence damping-off involving death of seedlings before they emerge from the soil. The seed may fail to germinate or the radicle of the seedling may be attacked as it develops.
- » Post-emergence damping-off involving death of the very young, succulent seedlings after they have emerged from the soil.
- » Infected portion shows water soaked lesions and browning of seedlings at collar region, root decay and death of seedlings in patches.
- » Water logging aggravates disease incidence.

- Top damping-off involving infection of cotyledons retained in the seed coat for extended periods, resulting in death of the seedling.

Oomycete:

- *Pythium* spp. (*Pythium deliense*, *Pythium intermedium*, *Pythium myriotylum*, *Pythium spinosum*, *Pythium mamillatum*, *Pythium ultimum*, *Pythium irregulare*, *Pythium paroecandrum* and *Pythium perplexum*) and
- *Phytophthora* spp. (*Phytophthora cryptogea, Phytophthora cambivora*)

Fungi:

- *Botrytis cinerea,*
- *Cylindrocladium* spp. (*Calonectria quinqueseptata*, *Calonectria ilicicola*, *Calonectria kyotensis*, *Cylindrocladium scoparium*, *Cylindrocarpon destructans*)
- *Alternaria alternata,*
- *Fusarium* spp. (*Fusarium avenaceum*, *Fusarium longipes*)
- *Thanatephorus cucumeris* and
- *Rhizoctonia solani*

Management

- Avoid water logging.
- Prepare raised beds.
- Use of pathogen-free seed.
- Direct sowing into conical plastic tubes, containing pathogen-free growing media.
- Thinning seedlings to one per tube as soon as possible (before 7 cm in height)
- Use of pathogen-free water.
- Culling seedlings for uniform height with removal of diseased or dead seedlings.
- Seed treatment with captan 2 g/kg seeds *P. fluorescens* 10 g/ kg seed.
- Avoid wounding plants as these will allow entry of the fungus.

- Adequate spacing to avoid crowding and the formation of high humidity.
- Do not overwater as disease is more severe in very wet soils.
- Remove affected seedlings.
- Apply *Pseudomonas fluorescens* 2.5 kg/ 50 kg FYM to nursery bed.
- Drench copper oxy chloride 0.25 % or carbendazim 0.1% or thiabendazole 0.1%, captan 0.2% and thiram 0.1%.

2. Vascular wilt

Symptoms

- The affected plants showed characteristic symptoms of wilting.
- Initially the lower foliage became flaccid and dried up.
- Slowly and gradually, the wilting proceeded upwards eventually killing the growing shoot.
- The wilted plants could be easily located in the plantations because of their dried up leaves.
- The vascular tissues of roots and stem of the diseased plants showed typical browning, abundant mycelium and spores characteristic of vascular wilt.

Fungus: *Fusarium oxysporum* f.sp. *eucalypti*

Management

- Soil drench and foliar spray with *Pseudomonas fluorescens* 2%.
- Solar heating (45°C–52°C) of moist soil under clear polythene sheeting (50 μm thick) during summer minimize the infection.

3. Cylindrocarpon **root rot**

Symptoms

- The initial symptom was wilting and drying up of leaves.
- The roots of the affected trees became discoloured and showed rotting.
- The infection generally spread up to the collar area and produced a canker.

- » The affected trees died within one to two months.
- » The leaves of the affected plants, especially the bottom ones, turned reddish purple in colour and dried up.
- » No wilting of foliage was observed.
- » The plants were killed within two to three weeks after the change in colour of the foliage was noticed.
- » The roots of diseased trees showed browning and rotting.

Fungus: *Cylindrocarpon lucidum* state of *Thelonectria lucida*

Management

- » The pathogen is a weak parasite; prevent the injury/wounds in any of the trees parts.

4. Woody stem and root rots

Symptoms

- » Root rot disease is characterized by expanding patches of dead and dying trees.
- » The foliage of affected trees usually becomes paler green and dull in appearance with reduced leaf size and eventual leaf shedding.
- » Crown dieback commences with flattening of the normally conical apex characteristic of rapidly growing Eucalyptus, and death of fine branches.
- » Epicormic shoots may be observed
- » Rotting of root predisposes trees to wind-throw.
- » Root rots of forest trees and plantations can be partly distinguished based on the appearance, especially colour, of infected roots and the fungal structures associated with them.

Fungi: Red root disease (associated with *Ganoderma lucidum; Ganoderma colossum*, brown root rot (caused by *Phellinus noxius*) and the white root rot (*Rigidoporus lignosus*).

- » Roots affected by red root rot (*Ganoderma* infection) are covered by a reddish-brown fungal crust, with white mycelia growth on underside of bark. *Ganoderma* spp., especially when associated with heart rot of large standing live or dead trees form conspicuous bracket shaped fruiting bodies in severely affected trees.

» Brown root rot (*P. noxius*) is a thick brown mycelia mantle on the surface of woody roots that often extends to the above-ground stem. These hyphal mats exude sticky fluid which causes soil particles to adhere. The fungus invades the main stem where it causes a white pocket rot with a characteristic honeycomb appearance.

» White root rot (*Rigidoporus lignosus*) is a very important disease of rubber, oil palm, *Eucalyptus* sp. However, white root rot does not appear to have been formally reported in eucalypts plantations.

» *Armillaria luteobubalina* is the most damaging of the four *Armillaria* species recorded on Eucalyptus.

Management

» Maintain the trees in healthy condition – application of manure.

» Do not wound the tree. If any wounds are there apply Bordeaux paste.

» Remove the sporocarp and destroy.

» Scrap the mycelia portion and apply Bordeaux paste.

» Remove severely affected trees.

» Isolate the affected tree from healthy trees.

» Drench the soil near the tree with 40 litres of 1% Bordeaux mixture at quarterly interval for thrice a year and repeat after 2-3 years.

» For *Ganoderma* / root rot apply root injection. In the infected plant, a downward hole is made to a depth of 3-4" at an angle of 45° at the height of 3' from the ground level with the help of an auger. The solution containing of Aureofungin sol 2 g and copper sulphate of 1 g in 100 ml of water is taken in a saline bottle and the bottle is tied with the tree. The hose is inserted into the hole and the stopper is adjusted to allow the solution in drops. After the treatment, the hole is covered with clay or do root feeding where the root region is exposed; actively growing young root is selected and given a slanting cut at the tip. The root is inserted into a polythene bag containing 100 ml of the tridemorph (0.1%) fungicidal solution (2 ml/ litre). The mouth of the bag is tied tightly with the root.

5. Leaf web blight

Symptoms

- Soil borne fungal disease.
- Leaves show water soaked greyish brown blotches.
- Infected leaflets often join together by the fungal hyphae - web blight.
- Leaflets or entire leaf become detached prematurely.
- Group blighting of seedlings in the nursery.

Fungus: *Rhizoctonia solani* (Teleomorph: *Thanatephorus cucumeris*)

Management

- Proper sanitation and weeding
- Foliar application of fungicide solution, triademefon 0.1% or carbendazim 0.1% at fortnightly intervals.
- Soil drenching with carbendazim, or quintozene followed by control of watering to prevent excessive soil moisture helps to check further spread of the disease.

6. Rust

Symptoms

- Initially, an egg yellow (orange) pustule (uredinia) occurs on young leaves in nursery and young plantations.
- Older pustules contain both single celled uredospore and two-celled, stalked teliospores.
- The rust attacks both foliage and young green shoots on which cankers are formed, with rupture of the epidermis and phloem.
- The presence of these cankers can cause stem distortion and crowns of badly affected trees become stunted and multi-stemmed.

Fungus: *Austropuccinia psidii* (Syn: *Puccinia psidii*)

Management

- Control can be readily achieved by use of species with a high or moderate level of resistance to rust and by selection within hybrid progeny for resistant clones.
- Spray with Mancozeb 0.2% or Triademenol 0.2% or Triforine 0.1% twice at 15 days interval.

7. Powdery mildew

Symptoms

- On leaves and young shoots, producing a thick layer of densely inter-woven white mycelium on the surface of leaves and shoots.
- Sometimes causing spotting and malformation of older growth.

Fungi:

- *Erysiphe cichoracearum* on *E. calycogona, E. camaldulensis, E. campaspe, E. cladocalyx, E. globulus, E. leucoxylon, E. leucoxylon* var. *rosea, E. microcarpa, E. pauciflora, E. pauciflora* ssp. *niphophila, E. platypus, E. polyanthemos, E. populnea, E. porosa, E. pulverulenta, E. tetragona, E. viminalis* and *E. viridis*
- *Erysiphe polyphaga* on *E. crebra* and *E. moluccana*
- *Sphaerotheca aphanis* on *E. albens, E. cinerea, E. crebra, E. diversicolor, E. grossa, E. megacarpa, E. nutans, E. paniculata, E. tereticornis and E. torquata*
- *Sphaerotheca macularis* on *E. algeriensis, C. citriodora, E. cornuta, E. diversicolor* and *E. gomphocephala;*
- *Sphaerotheca pannosa* on *E. albens, E. camaldulensis, E. gunnii, E. moluccana, E. nitens, E. perriniana* and *E. tereticornis;*
- *Oidium eucalypti* on *E. botryoides, E. camaldulensis, E. crenulata, E. globulus, E. globulus* ssp. *maidenii, E. gunnii, E. muellerana, E. perriniana, E. tereticornis* and *E. viminalis.*

Management

- Spray recommended doses of *Trichoderma harzianum, Ampelomyces quisqualis, Bacillus* and *Ulocladium.*

» Spray with Chlorothalonil 0.1% or Triademefon 0.2% or Maneb 0.2% or Zineb 0.2%.

8. Black mildew

Symptoms

» Species of *Meliola* grow on the surfaces of leaves and stems and form thick, black, radiate, velvety colonies of up to 1 cm diameter. In cases of heavy infestation, the entire leaf surface may be covered by the fungus.

» The infection is usually more frequent on upper than lower leaf surfaces. Sometimes young stems and twigs can also be infected.

» Numerous minute spherical fruiting bodies develop on the fungal thallus on the leaf surface.

» Derives nutrients through haustoria as a result the tree suffers from loss of chlorophyll and results in reduced soluble sugars, starch, total protein, amino acid.

Fungi: *Meliola densa, Meliola amphitricha*

Management

» Proper sanitation

» Spray fungicides and insecticides to eliminate scale insects and mealybugs.

9. Cryptosporiopsis leaf and shoot blight

Symptoms

» Infections develop on both leaves and shoots of eucalypts.

» Leaf spots occur on both sides of the leaves and vary in size, shape and colour, include large, brown, spreading necrotic lesions leading to a leaf blight symptom, circular or sub-circular spots 1–2 cm in diameter; irregular chocolate-brown or greyish spots covering much of the leaf area; irregular roughened or corky lesions with eruption and necrosis of epidermal tissue, sometimes localised along veins, on which the fungus fruits.

» Terminal shoots of young trees can be totally defoliated and are commonly blighted.

» Plantation trees become flattened, main stems suffer dieback and multiple branching, apical dominance is reduced, and growth can be stunted.

Fungus: *Cryptosporiopsis eucalypti*

Management

» Growing of resistant species like *Eucalyptus pellita* and moderately resistant species such as *E. camaldulensis* sp *simulata* and *E. tereticornis.*

» Soil drench and foliar spray with Carbendazim 0.1%.

10. Cylindrocladium foliar spot and foliar blight

Symptoms

» The disease first shows as greyish water-soaked spots on young leaves. These spots coalesce and develop into extensive necrotic areas.

» Large numbers of shining white spores can be seen at the margin of lesions, on older necrotic portions of leaves, especially along midribs on the abaxial surfaces, and on fine young shoot tips which are killed, resulting in leaf and shoot blight symptoms.

Fungi: *Calonectria hederae* (Syn: *Cylindrocladium reteaudii*),

Calonectria cylindrospora (Syn: *Cylindrocladium scoparium*),

Calonectria ilicicola (Syn: *Cylindrocladium parasiticum*),

Calonectria pyrochroa (Syn: *Cylindrocladium ilicicola*),

Calonectria ovata (Syn: *Cylindrocladium ovatum*),

Calonectria candelabra (Syn: *Cylindrocladium candelabrum*),

Calonectria brassicae (Syn: *Cylindrocladium gracile*)

Management

» Grow resistant species like *Eucalyptus pellita* and *Eucalyptus brassiana.*

» Collection and destruction of infected leaves.

» Maintain proper hygiene in the field.

» Provide proper spacing and removal of auxiliary branches for the entry of sunlight.

» Plough in between the tree rows to expose the soil borne chlamydospores.

» Foliar spray or soil drench with carbendazim 0.1%.

11. Cylindrocladium shoot/ seedling blight

Symptoms

» The infection affects both stem and foliage.

» The stem infection, observed in coppice shoots and branches of young trees, appeared somewhere on branches and caused canker characterized by a dull brown depression on the stem.

» Stem infection, which is more common in young plants and coppice shoots, is observed only during the monsoon.

» Stem infection coupled with severe leaf blight resulted in die-back of shoots.

» Severity of infection generally depended upon the microclimatic conditions, especially rainfall, humidity and also the age of the plants.

» During high humid periods, *Cylindrocladium* was seen to produce profuse mycelium and conidial mass.

» During the rainy season numerous fructifications (teleomorph) developed on the dead stem.

» Leaves of all maturities from young and old plants, epicormic and coppice shoots were found to be equally susceptible, the leaf infection appeared in the form of minute grayish-black spots which coalesced to form large necrotic area.

» In mature leaves occasionally the infection initiated either from leaf tip and spread downwards or from the margin and gradually spread towards the midrib.

» During the dry period, the lesions became dull pale brown.

» Severe infection of leaves, which caused blight, resulted in extensive premature defoliation.

Fungi: *Calonectria quinueseptata* and its anamorph *Cylindrocladium quinqueseptatum*,

Calonectria pyrochroa and its anamorph *Cylindrocladium ilicicola*,

Calonectria indusiata and its anamorph *Cylindrocladium theae*,

Cylindrocladium clavatum,

Calonectria morganii and its anamorph *Cylindrocladium scoparium.*

Management

- » Avoid prolonged wetting of nursery and main plantations.
- » Prophylactic spray with Carbendazim 0.2% or captan 0.25% of the planting stocks.

***12. Chrysoporthe* stem canker**

Symptoms

- » Initially slight sunken, elongate areas measuring about 15-20 cm on the trunk either at the base or above ground, just after the south-west monsoon.
- » Basal cankers, which can extend several metres up the stem and less severe cankers often associated with branch stubs.
- » Stem girdling, leads to wilting and death of trees suddenly during hot dry weather.
- » Severe bark cracking over brown necrotic sapwood.
- » Infected stems become discoloured, deep red (ruby color) or brown, through copious secretion of kino from cankers, which dries on the bark surface.
- » Large basal cankers can kill trees during the first 2–3 years of growth.
- » Compared with normal wood, cankered wood leads to loss in pulp yield.
- » During, dry season cankers develop on barks with necrotic tissues that develop pycnidia and perithecia.

Fungi: *Chrysoporthe cubensis* (Syn: *Cryphonectria cubensis*), *Diaporthe cubensis, Endothia gyrosa, Cryphonectria eucalypti, Endothia tropicalis*

Management

- » *E. grandis* × *E. urophylla* hybrids and selection within these hybrids for resistant genotypes.

- Clonal propagation of resistant genotypes has provided an effective control of the disease.
- Brush on application of Copper oxychloride 0.25% or Bordeaux mixture 1% on bark and cut surface of the stumps.

13. Amphilogia stem canker

Symptoms

- The infection usually occurs on lower part of the stem, often near the ground during the monsoon and by next few months depressed areas, 30-40 cm long on the bark become visible.
- The tissues beneath the canker turned necrotic and get killed.
- During the dry period splitting of the bark appeared on cankered areas.
- The pycnidia, orange-red in colour, developed over the bark arranged in vertical linear rows or scattered during the wet period.
- The pycnidia produce long orange yellow tendrils of spores during monsoon. Characteristic ascomata with log beaks developed during the following dry period on the cankered areas, either separately or interspersed with pycnidia.
- No gummosis was noticed on the cankers as in the case of *C. cubensis* canker.
- Infection also occurs on branches and twigs resulting in twig and branch die-back.

Fungi: *Amphilogia gyrosa* (Syn: *Cryphonrctria gyrosa*), *Endothia havanensis*

Management

- Drenching with Benomyl 0.1% or Carbendazim 0.1% or Tridemefon 0.1% or Mancozeb 0.5% or Chloroneb 0.5% or Triforine 0.1% or Thiabendazole 0.1%.

14. Coniothyrium stem canker

Symptoms

- Discrete black or dark brown, sunken lesions on young green tissue of trees.
- Lesions merge -necrotic patches on stem.
- Infections penetrate the cambium.

» Produces kino packets in wood.

» Severely cankered stems produce epicormic shoots around the cankers, growth stops- bushy appearance.

Fungus: *Coniothyrium zuluense*

Management

» Proper sanitation and weeding.

» Selection of resistant root stock.

***15. Cytospora* stem canker**

Symptoms

» Infection occurred on any part of the stem.

» The infected region showed browning and leaves wilted and defoliated.

» Numerous black conidiomata (pycnidia) developed scattered over the entire infected region.

» Complete girdling of the phloem and cambium usually resulted in death of the affected branches.

» Infection on the main stem of coppice shoots was initially observed at the base near the stump and later, it gradually spreads towards the apex.

» Girdling due to the canker at the base resulted in wilting of the leaves and death of shoots.

» In severe case, the infection even spreads to roots killing the stump.

» Generally, all the shoots of a stump got infected and died.

Fungi: *Cytospora eucalypticola* causes main stem canker and *Cytospora eucalypti* (*Valsa eucalypti*) causes twig and branch canker.

Management

» No control measures are recommended.

16. Valsa stem canker

Symptoms

- Infection is initiated at the base of the branch where a canker develops.
- Over the dead bark, numerous black fructifications are produced.
- Often more fructification is produced on the lower surface of a branch, away from the direct sunlight.
- The affected twigs and branches die due to complete girdling of phloem tissue.

Fungi: *Valsa eucalypti* and *Valsa eucalypticola*

Management

- No control measures are recommended.

17. Macrovalsaria stem canker

Symptoms

- Generally the infection occurs at the basal part of the stem, characterized by a large number of black fructifications, scattered over the bark.
- The affected stem develop canker and the underneath tissues show browning. When the stem is completely girdled, the foliage wilts and the tree slowly die.
- Epicormic shoots development does not occur.

Fungus: *Macrovalsaria megalospora*

Management

- No control measures are recommended.

18. Thyronectria stem canker

Symptoms

- Initially, the conidial state of the pathogen developed near the base of the stem which soon spread upwards and developed browning and when stem completely girdled, it brought about wilting and consequently death of tree.

- The perfect stage of the fungus was formed in loose clusters or scattered over the dead stem.
- No epicormic shoots developed on the affected trees.

Fungus: *Thyronectria pseudotricha*

Management

- The pathogen is a wound parasite, prevent the injury/wounds in any of the trees parts.

19. Hysterium stem canker

Symptoms

- The pathogen caused extensive cankers on branches and upper part of the stem.
- The affected tissues of the stem showed pronounced browning.
- On dead stem numerous characteristic black fructifications developed, aggregated or singly.

Fungus: *Hysterium angustatum*

Management

- No control measures are recommended.

20. Nattrassa stem canker

Symptoms

- The infection appeared on the stem near the ground and later spread upwards covering a large part of the stem.
- The infected area got differentiated in a depression, which later turned into a canker.
- The roots also got infected and complete girdling of the stem resulted in death of plants.
- The foliage gradually wilted and defoliated.
- On the canker numerous minute fructifications, pycnidia, of the fungus developed arranged in vertical broken lines.

- The tissue of the affected area showed grayish black discoloration in which dark brown, septate, thick-walled, inters and intra-mycelia were found in abundance.

Fungus: *Nattrassa toruloidea*

Management

- No control measures are recommended.

21. Pink disease

Symptoms

- Infection of young branches leads to killing and in old branches cankers are formed in barks with longitudinal cracking/ splitting.
- The apical shoot above the canker dies when the stem is completely girdled.
- Numerous epicormic shoots develop from the healthy stem just below the canker.
- The shoots also get infected and killed following wilting and drying up.
- These cankers serve as infection courts for heart rot fungi which cause heart wood decay and thereby wood gets disintegrated.
- Infection on stem and branches leads to girdling and death of parts above.
- Lower areas show epicormic branches with results in plant growth retardation.
- Later gum exudes from cankers.

Different stages of pink disease development: There are only four stages

i. **Pustules**: Appear on stem and branches within few weeks of infection at onset of monsoon and turn to pink to salmon colour bodies.

ii. **Cobweb stage**: Appear on the bark thin, white, arachnoid mycelia from pustule or from the edges of the pink encrustations during monsoon period.

iii. **Necator**: Conidial stage after the rains, orange-red in colour and borne mainly on the upper side of branches.

iv. **Pink encrustation**: Sexual stage of the fungus develops as thin light pinkish encrustation on the bark, under surface on the stem on the shady side.

Fungus: *Phanerochaete salmonicolor* (Syn: *Corticium salmonicolor*; *Erythricium salmonicolor*)

Management

- Remove affected portions and apply Bordeaux paste at cut ends.
- Clonal propagation of resistant clones.
- Do not injure the trees or make wounds.
- Spray with Carbendazim 0.1% at a weekly interval.

22. Lasiodiplodia stem (root collar) canker

Symptoms

- Stem cankers are dead areas of bark which sometimes extend into the underlying sapwood.
- Sunken lesions extend more than a metre along the branch or stem axis.
- Such 'diffuse' cankers may be darkly discoloured and cracked especially toward the centre. stems usually secrete kino from these lesions, discolouring the bark with reddish to dark brown pigments
- Branches and stems may be partially or completely girdled, causing crown dieback and possibly tree death.
- Fruiting bodies of the causal fungi can usually be found on the cankers are typically partially submerged in the outer bark but can be readily seen through a hand lens.
- Cankers are often associated with wounds or borer damage, or with branch stubs.
- Trees planted in unsuitable environments, for example infertile soils and climates to which they are poorly adapted (drought-prone areas), are more susceptible to canker diseases.
- Trees in dense stands, especially if suppressed, subjected to insect attack (either defoliation or stem borers), are also more readily invaded.
- Poor silvicultural operations resulting in stem wounds also predispose trees to infection.

Fungi: *Lasiodiplodia theobromae* (Syn: *Botriodiplodia theobromae*), *Dothiorella* sp. and *Fusicoccum* sp.

Management

- Provide proper spacing for light and aeration.
- Spray with Tebuconazole 0.1% or Carbendazim 0.5% or Thiobendazole 0.5%.

23. Seiridium stem canker

Symptoms

- Characteristic lenticular lesions which are normally sunken in the centre with some longitudinal cracking and with stem swellings around the lesions.
- Some lesions girdle more than half the stem circumference.
- Branches or stems are killed above the lesions.

Fungus: *Seiridium eucalypti*

Management

- Provide proper spacing for light and aeration.
- Spray with Carbendazim 0.5% or Thiobendazole 0.5%.

24. Mycosphaerella leaf spot

Symptoms

- Infected leaves develop spots and blotches, large lesions develop, often with crinkling of the leaves.
- Affected trees suffer premature defoliation and severe disease can cause stunting of trees.

Mycosphaerella cryptica:

- Reddish-brown, irregular-shaped spots with a prominent purple margin present on both sides of young, expanding leaves. Spots eventually turn dark grey. Fruiting bodies develop on both sides of leaves.
- Presence of necrotic spots or patches and pseudothecia on leaves.
- Infected leaves often crinkled and distorted and cast prematurely.

» Cankers (up to 25 mm long) on infected shoots.

» Gum exudation is often associated with canker development and dieback occurs if cankers girdle shoots.

Mycosphaerella nubilosa:

» Creamy yellow to pale brown, irregular-shaped spots found on both sides of young, expanding leaves.

» Spots later turn grey-black on the underside.

» Fruiting bodies develop on the underside of leaves only.

Fungi: *Mycosphaerella nubilosa, Mycosphaerella cryptica, Mycosphaerella marksii, Mycosphaerella parva, Mycosphaerella molleriana, Mycosphaerella madeira, Mycosphaerella readeriellophora, Mycosphaerella communis, Mycosphaerella lateralis, Mycosphaerella aurantia.*

Management

» Foliar spray with Carbendazim 0.1% or Mancozeb 0.2% or Chlorothalonil 0.1% at 15 days interval.

25. Aulographina leaf spot (Target spot)

Symptoms

» Initially small, circular, brown, roughly circular (2-15 mm in diameter) spots with a raised, crusty centre, and a dark margin on both sides of leaves and petioles, shield-like fruiting bodies of the asexual state develop on the surface of the lesion.

» These are followed by black, elongated fruiting bodies of the sexual state which are often branched and open by splitting along their length.

» Minute (up to 2 mm long), black flecks (the fruiting bodies of the fungus) scattered on the leaf spots.

» Spots on one side of the leaf are rarely visible on the other side.

Fungus: *Aulographina eucalypti* teleomorph of *Thyrinula eucalypti*

Management

» No control measures are recommended.

26. Pseudocercospora leaf spot

Symptoms

- Sub-circular to angular-shaped, pale to dark brown spots with a brown to purplish margin which may be somewhat raised.
- Lesions becoming greyish-brown to grey when old.
- Lesions found singly or merging together on both the upper and lower surfaces of the leaf.
- Individual lesions range in size from 1-12 mm in length.

Fungi:

- *Pseudocercospora acerosa* on *Eucalyptus baxteri, E. nitens, E. verrucata*
- *Pseudocercospora crousii* on *Eucalyptus dendromorpha, E. fastigata, E. muelleriana, E. pilularis, E. regnans, E. stenostoma*
- *Pseudocercospora eucalyptorum* (Syn: *Cercospora eucalyptii* sensu lacto) on *Eucalyptus botryoides, E delegatensis, E. dendromorpha, E. fastigata, E. fraxinoides, E. globoidea , E. globulus, E. nicholii, E. nitens, E. obliqua, E. ovata, E. radiata, E. regnans, E. resinifera*
- *Pseudocercospora subulata* on *Eucalyptus nitens*

Management

- No control measures are recommended.

27. Sonderhenia leaf spot

Symptoms

- Small (1-3 mm in diameter) spots with a distinct purple-red margin scattered over both upper and lower leaf surfaces.
- Spot surfaces are often raised.
- The two species of Sonderhenia form lesions that are very similar in appearance and cannot readily be distinguished.
- The species are separated microscopically on spore dimensions and they also differ in host association.

Fungi:

- *Sonderhenia eucalypticola* (sexual) stage: *Mycosphaerella walkeri* on *Eucalyptus fraxinoides, E. globulus* subsp. *globulus, E. nicholii, E. nitens, E. sideroxylon.*
- *Sonderhenia eucalyptorum* (sexual) stage on *Mycosphaerella swartii* on *Eucalyptus delegatensis, E. elata, E. fastigata, E. fraxinoides, E. globoidea, E. johnstonii, E. leucoxylon, E maculata, E. muelleriana, E. obliqua, E. regnans, E. viminalis.*

Management

- No control measures are recommended.

28. Septoria leaf blight

Symptoms

- Pale yellow spots on both sides of the leaf.
- Spot colour soon changes to a deep carmine-red and finally to brown with a residual carmine margin.
- Lesions regular in shape, often merging together and covering large areas of the leaf.
- Exuded spore masses visible on the lesions as minute dark-coloured spots.

Fungi: *Phaeophleospora eucalypti, Phaeophleospora epicoccoides, Phaeophleospora lilianiae, Phaeophleospora destructans*

1. ***Phaeophleospora epicoccoides***: *Kirramyces epicoccoides, Cercospora epicoccoides, Hendersonia grandispora, Phaeoseptoria eucalypti, Phaeoseptoria luzonensis*
2. ***Phaeophleospora eucalypti***: *Kirramyces eucalypti, Cercospora eucalypti, Pseudocercospora eucalypti, Septoria pulcherrima, Stagonospora pulcherrima*
3. ***Phaeophleospora destructans***: *Phaeophleospora destructans, Kirramyces destructans*

Management

- Removal of affected leaves and dried debris.

29. Phaeoseptoria leaf spot

Symptoms

- Initially appeared on mature leaves as purple to brownish purple amphigenous spots which were characteristically angular and marked by veins.
- All the mature leaves had defoliated prematurely due to heavy infection.
- Later, the spot turned necrotic, minute black fruiting bodies (pycnidia), generally more on the abaxial surface, developed embedded in the leaf tissue.
- Pycnidia produced long grayish-back tendrils which appear as brownish black wooly mass on both the leaf surfaces.
- Due to rain or dew, the conidia got dispersed from the tendrils and formed a black layer over the leaf surface.

Fungus: *Phaeoseptoria eucalypti*

Management

- Spray carbendazim 0.03% or thiobendazole 0.1% or benomyl 0.2% at weekly interval.

30. Coniella leaf blight

Symptoms

Coniella fragariae

- Infection occurs in the form of circular lesions, which spread to the whole leaf and became necrotic and often caused foliage blight.
- Numerous pale to dark brown coloured, immersed to semi-immersed pycnidia of the fungus arranged more or less in concentric rings develop over the necrotic lesions.
- As the necrotic lesions get enlarged, new rings of pycnidia develop.
- During the wet period, pale olive green to pale brown conidial ooze developed which dispersed by rain splashes and the inoculum spreads to the new healthy leaves.

Coniella australensis

- Pycnidia developed irregularly over the necrotic lesions.
- During the wet period, pale olive green to pale brown conidial ooze developed which dispersed by rain splashes and the inoculum spreads to the new healthy leaves.

Fungi: *Coniella australiensis, Coniella castaneicola, Coniella fragariae, Coniella granati, Coniella minima*

Management

» Spray mancozeb 0.2% at weekly interval.

***31. Guignardia* leaf spot**

Symptoms

» Initially the symptoms developed as minute purplish spots on the upper surface of the leaves.

» Later, they got enlarged and became purplish brown necrotic areas, 4-5 mm across, angular to irregular, often lined with veins.

Fungus: *Guignardia citricarpa*

Management

» No control measures are recommended.

***32. Trimmatostroma* leaf spot**

Symptoms

» Brown, roughly circular (2 - 18 mm in diameter) spots, frequently composed of concentric rings of different shades of brown.

» Lesion centre often raised and with a crusty appearance.

» Small, black dots (the fruiting structures) visible on the spots.

Fungi:

» *Trimmatostroma bifarium* on *Eucalyptus delegatensis, E. fastigata, E. fraxinoides, E. nitens, E. obliqua, E. regnans, E. sieberi.*

» *Trimmatostroma excentricum* on *Eucalyptus delegatensis, E. fastigata, E. obliqua, E. pauciflora subsp. niphophila, E. regnans, E. sieberi.*

Management

- No control measures are recommended.

33. Bacterial wilt

Symptoms

- Symptoms develop shortly after planting.
- Affected trees show wilting, leaf drop, stem death and reduced growth rate.
- Vascular discoloration (black) commonly occurs, gummy ooze, roots die and basal cankers may be found on affected trees.
- Trees usually wilt and die.
- Stem sections, cut through discoloured vascular tissue exude bacterial masses if incubated moist in plastic bags for 24 hours or suspended in water.

Bacterium: *Ralstonia solanacearum*

Management

- Removal of affected plants and dried debris.
- Drenching with Copper oxy chloride 0.25%.

34. Bacterial leaf blight

Symptoms

- The disease symptoms may vary depending on leaf age, development stage of the lesion and species of Eucalyptus, but in general they are characterized by water soaked, angular, interveinal, and anphighenous lesions.
- Lesions evolve into necrotic brown to straw-colored lesions, sometimes with chlorotic or reddish edges, distributed over the limb or just on one half of the leaf blade, but especially along the main vein and on the edges of the leaf.
- Deformities in the leaf limb and abortion of the lesioned area are usually observed, resulting in cut or perforated limbs.

- Defoliation commonly occurs on highly susceptible genotypes due to the early senescence of infected leaves.

Bacterium: *Xanthomonas campestris* pv. *eucalypti*

Management

- Removal of affected plants and dried debris.
- Drenching with Copper oxychloride 0.25%.

35. Little leaf blight

Symptoms

- The affected plants showed prominent stunting and produced much smaller leaves when compared to healthy ones.
- The new leaves showed considerable reduction in size and became thin, pale, scaly with narrow lamina.
- The apices of such leaves often showed browning.
- The internodes became stunted and all axillary buds sprouted resulting in bushy shoots with abnormal minute leaves.
- Affected trees became weak due to reduction in stem girth and height growth.

Bacterium: *Candidatus* Phytoplasma

Management

- Removal of affected plants and dried debris.
- Spraying of tetracycline minimize the symptom expression.

36. Mosaic

Symptoms

- All the leaves young as well as mature, of the affected plants had mosaic symptoms; i.e., light pale to yellowish white, irregular patches.
- Even the new leaves showed these symptoms.

» The diseased plants did not show any effect on growth, except that affected leaves became leathery and thick.

Pathogen: Virus

Management

» Removal of affected plants and dried debris.

Other Diseases

37. **Root rot**: *Phytophthora cinnamomi, Armillaria mellea, Botryosphaeria ribis, Botryosphaeria berengeriana, Clitocybe tabascens, Helicobasidium compactum, Polyporus dryadeus.*

38. **Damping off and seedling blight**: *Cylindrocladiella parva*

39. **Seedling and cutting wilt**: *Cylindrocladium floridanum*

40. **Leaf spot on *Eucalyptus tereticornis***: *Ramularia pitereka, Alternaria alternata, Readeriella mirabilis, Rytisma eucalypti, Cylindrocladium quinqueseptatum, Colletotrichum gloeosporioides, Cylindrocladium curvatum, Cylindrocladium floridanum, Cylindrocladium ilicicola, Cylindrocladium scoparium, Cylindrocladiella camelliae, Cylindrocladiella peruviana, Phomopsis* sp., *Kirramyces eucalypti*

41. **Leaf spot on *Eucalyptus citriodora***: *Coniella minima, Phomopsis* sp., *Marsonina* sp. *Cylindrocladium scoparium.*

42. **Leaf spot on *Eucalyptus grandis***: *Coniella fragariae, Cylindrocladium quinqueseptatum, Cylindrocladium curvatum, Cylindrocladium floridanum, Cylindrocladium ilicicola, Cylindrocladium pteridis, Cylindrocladium scoparium, Cylindrocladium spathulatum, Cylindrocladium theae, Cylindrocladiella camelliae.*

43. **Leaf spot on *Eucalyptus robusta***: *Readeriella mirabilis, Harknessia hawaiiensis, Harknessia insueta, Colletrotrichum gloeosporioides, Cylindrocladium ovatum* sp. *novus.*

44. **Necrotic spot**: *Ceuthospora innumera, Piggotia substellata.*

45. **Nursery infection**: *Rhizoctonia solani, Alternaria tenuis, Alternaria tenuissima, Calonectria theae, Cercospora epicoccoides, Cytospora eucalypticola.*

46. **Defoliation**: *Harknessia eucalyptina*

47. **Canker**: *Hypoxylon annulatum, Hypoxylon howeianum, Poria epimiltina, Lophoaria cinerascens, Phellinus gilvus, Mycoacia uda, Botryosphaeria ribis.*

48. **Heart rot**: *Fistulinna hepatica, Fomes fomentarius, Fomes robiniae, Fomes robustus, Ganoderma applanatum, Pleurotus lampas, Polyporus portentosus, Polyporus sulphureus.*

49. **Gummosis and trunk rot**: *Phaeolus schweinitzii* and *Fomes* sp.

50. **Crown gall**: *Rhizobium radiobacter*

51. **Die-back**: *Xanthomonas eucalypti*

Selected References

Blum LEB, Dianese JC, Costa CL. 1992. Comparative pathology of *Cylindrocladium clavatum* and *C. scoparium* on Eucalyptus spp. and screening of Eucalyptus provenances for resistance to *Cylindrocladium* damping-off. Tropical Pest Management. 38, 2155–2159.

Braun U, Dick MA. 2002. Leaf spot diseases of eucalypts in New Zealand caused by *Pseudocercospora* species. New Zealand Journal of Forestry Science. 32(2): 221-234.

Brown, B.N. (eds). Diseases and pathogens of eucalypts, 293–306. CSIRO, Collingwood, Victoria.

Coutinho T, Preisig O, Mergaert J, Cnockaert MC, Riedel KH, Swuings J, Wingfield MJ. 2002. Bacterial blight and dieback of *Eucalyptus* species, hybrids, and clones in South Africa. Plant Disease 86:20-25.

Coutinho TA, Wingfield MJ, Alfenas AC, Crous PW. 1998. Eucalyptus rust: a disease with the potential for serious international implications. Plant Disease, 82(7), 819-825.

Dick M. 1990: Leaf-inhabiting fungi of eucalypts in New Zealand. II. New Zealand Journal of Forestry Science 20: 65-74.

Gadgil PD. 2005: Fungi on trees and shrubs in New Zealand. Fungi of New Zealand Volume 4. Fungal Diversity Research Series 16: 1-437.

Hood IA, Gardner JF, Kimberley MO, Molony K. 2002. Variation among eucalypt species in early susceptibility to the leaf spot fungi *Phaeophleospora eucalypti* and *Mycosphaerella* spp. New Zealand Journal of Forestry Science 32, 235-255.

Kile GA. 2000. Woody root rots of eucalypts. In: Keane, P.J., Kile, G.A., Podger, F.D. and pathogen of Eucalyptus in Indonesia. South African Journal of Botany 62:325–327.

Sharma JK, Mohanan C, Florence EJM. 1984. Nursery diseases of Eucalyptus in Kerala. European Journal of Forest Pathology 14:77–89.

Sharma JK, Sankaran KV. 1991. Epidemiological studies of *Rhizoctonia* web blight of *Albizia falcataria*. Indian Phytopathology 44:201–205.

Tamura T, Hayakawa M, Hatano K. A new genus of the order Actinomycetales, *Cryptosporangium* gen. nov., with descriptions of *Cryptosporangium arvum* sp. nov. and *Cryptosporangium japonicum* sp. nov. Int J Syst Bacteriol 1998;48:995–1005.

Thiribhuvanamala G. 2017. Major diseases affecting pulp wood trees and their management. In: K.T. Parthiban, R. Seenivasan (Eds) Plantation and Agroforestry Pulpwood Value Chain Approach, Scientific Publisher, India

Vaartaja O, Morgan GA. 1961. Damping-off etiology especially in forest nurseries. Phytopathology 51:35–42.

24 Diseases of *Ficus benghalensis* (Banyan)

S. No.	Disease	Pathogen
1.	*Septoria* leaf spot	*Septoria arcuata*
2.	*Alternaria* leaf spot	*Alternaria alternata*
3.	*Glomerella* leaf spot	*Colletotrichum* state *Glomerella cingulata*
4.	*Botryodiplodia* leaf spot	*Botryodiplodia theobromae*

1. Septoria leaf spot

Symptoms

- » The leaves of infected plants have dark black spot on lower surface of leaves.
- » Spores were found are yellowish long rod shaped one celled.
- » In growing stage the cells divide and forming 2-3 celled club shaped structure.
- » After some time of infection the infected area of leaves become dead and discarded from leaves.

Fungus: *Septoria arcuata*

Management

- » Spray with Carbendazim 0.1% or Mancozeb 0.1%.

2. *Alternaria* leaf spot

Symptoms

» The leaves of infected plants have brown, irregular, 3-10 mm necrotic spots with darker margins and yellow halos.

» Later, the spots may colease eachother and cause complete defoliation.

Fungus: *Alternaria alternata*

Management

» Spray with Carbendazim 0.1% or Mancozeb 0.1% or Chlorothalanil 0.2%.

3. *Glomerella* leaf spot

Symptoms

» Diseased spots were amphigenous, distinct, angular becoming irregular, sometimes coalescing to form large spots.

» The colour of the spots was light to dark brown.

Fungus: *Colletotrichum* state *Glomerella cingulata*

Management

» Spray with Carbendazim + Mancozeb 0.1% or Chlorothalanil 0.2% or Benomyl 0.1%.

4. *Botryodiplodia* leaf spot

Symptoms

» The disease starts along the margin and tip of the leaf and at later stages it encircles the whole leaf.

» Colour of the leaf spot was light brown with reddish tinge along the margins of the spot.

» These were amphigenous.

» Infected leaves become chlorotic, when a major portion has been infected, the leaves fall off.

Fungus: *Botryodiplodia theobromae*

Management

» Spray with Propiconazole 0.1% or Tebuconazole 0.1% or Chlorothalanil 0.2%.

Other Diseases

5. ***Bartalinia* leaf spot**: *Bartalinia robillardes*
6. ***Chiochorella* leaf spot**: *Chiochorella mangiferae*
7. ***Helminthosporium* leaf spot**: *Helminthosporium velutinum*
8. ***Phomopsis* leaf spot**: *Phomopsis bakeri*
9. ***Coniothyrium* leaf spot**: *Coniothyrium olivaceum*
10. **Rust**: *Exosporium fici*

Selected References

Firdousi SA, Rai AN, Vyas KM. 1990. New leaf spot disease of *Ficus benghalensis* L. caused by *Coniothyrium olivaceum* Bonard from India, Acta Botanica Indica, 18, 150-155.

Pathak H, Maru S, Satya HN and Silawat SC. 2015. Fungal Diseases of Trees in Forest Nurseries of Indore, India. Journal of Plant Pathology and Microbiology. 6:8

25 Diseases of *Ficus racemosa* (Cluster fig)

S. No.	Disease	Pathogen
1.	*Phyllosticta* leaf spot	*Phyllosticta* sp.
2.	Rust	*Cerotelium fici*

1. *Phyllosticta* leaf spot

Symptoms

- » Initially the lesions are circular or oval usually less than 5 mm.
- » Dark reddish brown lesions starting at the leaf margin and progressing inward.
- » As the disease progresses, lesions became irregular and when severe infection occurs, several spots coalesced to each other and cover large portions on the leaf, which gave a blighted appearance.

Fungus: *Phyllosticta* sp.

Management

- » Spray with Carbendazim 0.1% or Mancozeb 0.1% or Chlorothalanil 0.2%.

2. Rust

Symptoms

- It causes small dark brown spots on the lower surface of leaves.
- Fruiting body presented on lower surface of leaves.
- The spores have single celled, rounded and spiny outer layer.

Fungus*: Cerotelium fici*

Management

- Spray with Carbendazim 0.1% or Carboxin 0.1% or Triadimefon 0.2%.

Selected References

Jamaluddin, Goswami MG, Ojha BM. 2004 Fungi of India. Scientific Publishers (India) 112-114.

Sharma P, Singh N, Verma OP. 2011. First report of leaf blight on *Ficus religiosa* caused by *Phyllosticta* sp. J Plant Pathol Microbiol 2:106.

Pathak H, Maru S, Satya HN and Silawat SC. 2015. Fungal Diseases of Trees in Forest Nurseries of Indore, India. Journal of Plant Pathology and Microbiology. 6:8

26 Diseases of *Gmelina arborea* (Candahar tree)

S. No.	Disease	Pathogen
1.	Damping-off	*Pythium aphanidermatum*
2.	Foot rot/ wilt	*Fusarium oxysporum*
3.	*Poria* root rot	*Poria rhizomorpha*
4.	Root rot and collar rot	*Athelia rolfsii*
5.	*Phoma* stem rot	*Phoma nebulosa*
6.	Leaf and shoot blight	*Glomerella cingulata*, *Fusarium solani*
7.	Powdery mildew	*Phyllactinia gmelinae*
8.	Stem canker	*Thyronectria pseudotricha*, *Hendersonula toruloidea*
9.	*Phomopsis* leaf spot	*Phomopsis micheliae*
10.	*Rhizoctonia* leaf spot	*Rhizoctonia bataticola*
11.	*Phomopsis* canker	*Phomopsis gmelinae*
12.	*Pseudocercospora* leaf spot	*Pseudocercospora ranjita*
13.	*Colletotrichum* leaf spot	*Glomerella cingulata*
14.	*Corynespora* leaf spot	*Corynespora cassiicola*
15.	Stem canker and dieback	*Griphosphaeria gmelinae*
16.	Stem decay	*Lentinus squarrosulus*
17.	Stem rot	*Sclerotium* sp.

1. Damping off

Symptoms

- It causes pre emergence damping off where seeds rot before emergence.
- It causes post emergence damping off where seedlings rot after they emerge.
- Water soaked lesions and browning at seedlings at collar region, root decay.
- Death of seedlings in patches.
- Water logging aggravates disease incidence.

Oomycete: *Pythium aphanidermatum*

Management

- Avoid water logging and prepare raised beds.
- Seed treatment with *Pseudomonas fluorescens* at 10 g/ kg seed.
- Remove affected seedlings.
- Apply *Pseudomonas fluorescens* at 2.5 kg/ 50 kg FYM to nursery bed.
- Drench Copper oxychloride at 0.05%.

2. Foot rot/ wilt

Symptoms

- Soil borne fungal disease and systemic disease.
- Water-soaked depression on root portion.
- Roots turn dark brown.
- Yellowing of leaves and drooping of leaves.
- Wilting of plants leads to death of plants.
- Infected Plant hard is to pull from soil.
- When root portion is cut transversely, pinkish discolouration inside root tissues

Fungus: *Fusarium oxysporum*

Management

- Remove affected seedlings / plants.
- Soil drenching with Carbendazim 0.1% at monthly intervals.
- Application of *Trichoderma viride* at 2.5 kg/ 50 kg FYM and applied to plants.

3. *Poria* root-rot

Symptoms

- The severity of the disease has been noticed more on clayey sites where the soil becomes periodically waterlogged.
- The fungus perennates in the forest on woody debris in the soil and humus from which it spreads to the host through white coloured cord-like rhizomorphs often to a distance of 30 meteres, disseminating the disease from one tree to another.
- It causes brown cuboidal rot in the root resulting in death of root cambium followed by die back and death of affected trees both in natural regeneration and plantations.
- The sporophores of the fungus effused thin, brittle, poroid, crust, inseparable from the substrate, or white to pinkish cinnamon, pores round to angular, basidiospores hyaline, ellipsoid in shape.

Fungus: *Poria rhizomorpha*

Management

- Remove affected seedlings / plants.
- Soil drenching with Carbendazim 0.1% at monthly intervals.
- Application of *Trichoderma viride* at 2.5 kg/ 50 kg FYM and applied to plants.

4. Root-rot and collar rot

Symptoms

- Leaf shedding and subsequent death of the seedlings.

- » Premature defoliation noticed.
- » Infected collar region white mycelial growth is observed.
- » Later brown sclerotial bodies are seen.
- » Rotting of roots.

Fungus: *Athelia rolfsii* (Syn: *Sclerotium rolfsii*)

Management

- » Fumigation / soil solarization of nursery bed.
- » Remove affected seedlings / plants.
- » Application of *Trichoderma viride* at 2.5 kg/ 50 kg FYM and applied to plants.
- » Soil drenching with Carbendazim 0.1% at monthly intervals.

5. Phoma stem rot

Symptoms

- » Initially light brown flecks appeared on lower part of the stem, which soon changed to dark brown discolouration often extending upto the entire length of seedling.
- » Leaves show drying, seedlings wilt and eventually die.
- » At this stage the seedling showed wilting with flaccid and droopy leaves.
- » The affected seedlings failed to survive.
- » Numerous pycnidia developed on the dead stem, which produced spore mass in creamy white ooze from them on maturity.
- » Disease is severe under warm and humid conditions especially in overcrowded seedlings.

Fungus: *Phoma nebulosa*

Management

- » Removing the affected seedlings from the seed beds.

- Soil drench of Carbendazim 0.1% drench at weekly intervals.
- Soil application of *Pseudomonas fluorescens* at 2.5 kg/50 kg FYM and apply to beds at 500 g/bed.

6. Leaf and shoot blight

Symptoms

- The stem of the affected seedlings showed prominent dark brown discolouration near the tip which extended under high humidity upto the ground level and also spread to petioles and leaves.
- The infection of stem and leaves resulted in blighting of shoots.
- Subsequently the stem and leaves of diseased seedlings were decayed on which a large number of acervuli developed producing abundant conidia in pink gelatinous mass.
- The infected seedlings died outright.
- The disease spreads rapidly after initial appearance in patches, causing large scale mortality of nursery stock.
- Subsequent colonization by *Fusarium solani* hastens blighting.

Fungus: *Colletotrichum* state of *Glomerella cingulata* in association with *Fusarium solani.*

Management

- Spray with Carbendazim 0.1% or Mancozeb 0.1%.

7. Powdery mildew

Symptoms

- The pathogen produces infection spots on the lower surface of the leaves with corresponding pale yellow colour on the upper surface

Fungus: *Phyllactinia gmelinae (Syn: Phyllactinia suffulta* var. *gmelina)*

Management

- Spray with Wettable sulphur 0.1% or Triadimefon 0.1%.

8. Stem canker

Symptoms

- » Usually infection occurred on one side of the stem as a depression on which numerous anamorphs were produced.
- » The tissue of the depressed area developed into a canker.
- » As the fungus invaded surrounding healthy tissues, the canker increased in length and width, increase in size due to fungal invasion to the surrounding healthy tissues.
- » The bark later splits and peels-off exposing the dead wood.

Fungi: *Thyronectria pseudotricha* and *Hendersonula toruloidea*

Management

- » Prophylactic sprays with Carbendazim + Mancozeb 0.2%.
- » Wound dressing with Bordeaux mixture 1%.

9. Phomopsis leaf spot

Symptoms

- » Initial symptoms were small brownish spots, 2-3×5 mm dia, which later spread to form large light brown necrotic areas with dark brown centres and yellowish margins.
- » Gradually, these necrotic spots increased in size, occupying a large area of the leaf, and caused premature defoliation.

Fungus: *Phomopsis micheliae*

Management

- » Spray with Carbendazim 0.1% or Mancozeb 0.1%.

10. Rhizoctonia leaf spot

Symptoms

- » In the nursery, the spots were small, brownish, and more or less circular on the leaf blade.
- » The spots enlarged with the maturity of leaves.

- These were irregular in shape and scattered all over the leaf surface.
- The centre of the infected area was dark brown and surrounded by yellow halo.
- The leaves had more infection than, matured one.

Fungus: *Rhizoctonia bataticola*

Management

- Spray with Carbendazim + Mancozeb 0.2% or Thiophanate methyl 0.2%.

***11. Phomopsis* canker**

Symptoms

- The primary symptom was the appearance of a prominent sunken area, dark in colour, on the stem.
- In this area numerous irregular crevices running vertically also developed.
- The tissues beneath the bark were necrotic and discoloured.
- As a result of canker development, the affected trees showed dieback symptoms, characterized by yellowing and thinning of leaves; defoliation and death of the branches followed.
- Gradually, the canker spread longitudinally in both the directions as well as transversely, which resulted in complete girdling of the stem and drying up of the main shoot.
- Development of epicormic shoots, just below the canker was observed in all the affected trees

Fungus: *Phomopsis gmelinae*

Management

- Spray with Bordeaux mixture 1%.

***12. Pseudocercospora* leaf spot**

Symptoms

- Brownish-black circular to irregular spots bearing numerous conidia were confined initially on the abaxial surface.

- Due to profuse sporulation the spots resembled sooty mould.
- On the adaxial surface the spots appeared as light yellowish green patches.
- Smaller spots coalesced to form larger ones.
- In severe cases the spots were also developed on the adaxial surface and petioles.
- Severely infected leaves turned yellow and fell prematurely.

Fungus: *Pseudocercospora ranjita*

Management

- Spray with Carbendazim + Mancozeb 0.2%.

13. Colletotrichum leaf spot

Symptoms

- Initially, brown small spots, 1-2 mm across, surrounded by light yellow halo appeared on lower mature leaves.
- Under high humidity the spots enlarged and became round to angular in outline, with light and dark brown concentric bands.
- Severely infected leaves turned yellow and defoliated prematurely.
- Gradually the upper leaves also got infected and defoliated, thus leaving the bare stem with a few apical juvenile leaves.

Fungus: *Colletotrichum* state of *Glomerella cingulata*

Management

- Spray with Carbendazim 0.02% or Mancozeb 0.05% in nursery.

14. Corynespora leaf spot

Symptoms

- The initial symptom was the appearance of irregular pale green water-soaked lesions which soon turned dark brownish-black spots with light brown centres.

» Later these spots enlarged and coalesced to give rise to large necrotic area which led to premature defoliation.

Fungus: *Corynespora cassiicola*

Management

» Spray with Carbendazim 0.02% or Mancozeb 0.05% in nursery.

15. **Stem canker and dieback**

Symptoms

» Initially, the tender branches on top of the tree started dying from tip downwards. Gradually the infection spread to main stem killing it outright.

» As a result of die-back some epicormic shoots developed from the basal part of the stem.

» But these shoots also died because of downward spread of infection in the stem.

» The wood of the affected trees showed light to dark brown discolouration.

» Numerous black coloured ascomata developed on the dead branches.

Fungus: *Griphosphaeria gmelinae*

Management

» Spray with Carbendazim 0.2% or Mancozeb 0.1%.

16. **Stem decay**

Symptoms

» Decay started on one side of the stem at the tree and proceeded upwards.

» The affected area became soft and pulpy.

» During the rainy season sporophores of the fungus were produced in clusters throughout the decayed area, but more at the base.

Fungus: *Lentinus squarrosulus*

Management

» Diseased tree can be managed by following proper hygiene practices.

17. **Stem rot**

Symptoms

» The infection occurred at the base of the stem in the form of profuse mycelial growth accompanied with the formation of numerous fructifications.

» The affected part of the stem became soft and showed sign of rotting.

» The surrounding healthy tissues produced epicormic shoots

Fungus: *Sclerotium* sp.

Management

» Diseased tree can be managed by following proper hygiene practices.

Other Diseases

18. **Stem canker and vascular wilt**: *Ceratocystis fimbriata*

19. **Top dying**: *Nattrassia mangiferae* following insect (*Tingis beesoni*), *Botryosphaeria rhodina* (*Lasiodiplodia theobromae*), *Poria* sp.

20. **Pink disease**: *Phanerochaete salmonicolor*

21. **Stem gall**: *Agrobacterium tumefaciens*

22. **Leaf spot and die-back:** *Deptoshaeria gmelinae*

23. **Leaf spot**: *Macrophomina phaseolina, Phoma tropica, Cercospora volkameriae, Guignardia gmelinae, Phyllosticta gmelina, Gnomonia* sp., *Ascochyta* sp. and *Mycosphaerella* spp.

24. **Leaf blight**: *Alternaria alternata*

25. **Root rot:** *Gibberella fujikuroi, Athelia rolfsii, Thanatephorus cucumeris, Chaetophoma* sp., *Polyporus* sp., *Armillaria mellea, Ceratocystis fimbriata.*

26. **Collar-rot and wilt:** *Colletotrichum gloeosporioides* (Anamorphic *Glomerella*), *Thanatephorus cucumeris* (Anamorphic *Ceratobasidium*) and *Pythium* spp.

Selected References

Browne FG. 1968. Pests and Diseases of Forest Plantation Trees. Clarendon Press, Oxford, UK.

Florence EJM, Sharma JK. 1987. *Corynespora cassiicola*: a new leaf pathogen for *Gmelina arborea* in India. J. Tropical For. 3: 181–182.

Maringoni AC, Furtado EL. 1997. Pathogenicity of *Agrobacterium tumefaciens* on Gmelina. Summa Phytopatol. 23: 248–251.

Moldenke HN. 1984. Notes on the genus Gmelina (Verbenaceae). Phytologia 55: 308–342.

Muchovej JJ, Albuquerque FC, Ribeiro GT. 1978. *Gmelina arborea*–a new host for *Ceratocystis fimbriata*. Plant Dis. Rep. 62: 717–719.

Singh 2002. Nursery Diseases of *Dipterocarpus macrocarpus* and *Gmelina arborea* in Assam - New Records. The Indian Forester, 128, 25-32.

Wingfield M, Robison D. 2004. Diseases and insect pests of *Gmelina arborea*: real threats and real opportunities. New Forests 28, 227-243.

27 Diseases of *Grevillea robusta* (Silver oak)

S. No.	Disease	Pathogen
1.	Grey leaf spot	*Botryodiplodia theobromae*
2.	Leaf spot	*Cercospora agharkeri*
3.	Flowering parasite	*Dendrophthoe falcata*

1. Grey leaf spot

Symptoms

- Minute dull black spot appear on the leaves which later enlarge to form a grey spot with a dark margin.
- Several such spots join to form necrotic lesions on the lamina.
- A large number of minute black dot like asexual fruiting bodies called pycnidia are produced in the advanced stages of infection.

Fungus: *Botryodiplodia theobromae*

Management

- Destruction of infected leaf debris
- Spray with Carbendazim 0.1% or Chlorothalonil 0.2% or Mancozeb 0.2% as twice at 15 days interval.

2. Leaf spot

Symptoms

- » Small circular spot with a dark brown margin develops on old leaves.
- » As the disease advances, the centre of spot become white, thin and drops off leaving a shot holes.
- » In severe infection defoliation also occurs.

Fungus: *Cercospora agharkeri*

Management

- » Spray with Carbendazim 0.1% or Mancozeb 0.2% as twice at 15 days interval.

3. Flowering parasite (Giant mistletoe)

Symptoms

- » The parasite is a partial parasite.
- » It produces a very hard flattened prostrating stem which runs over the stem portions of the host and produce haustorial connections to drive the food source.
- » The point of parasitism becomes thickened to form a gall like structure from which multiple shootings ranging from 15-20 arise as a clusters.
- » The affected branch starts withering; drying from tip and dries off and produces erect, clustered branches from the galled portion.
- » The parasite is a shrubby plant with green leaves.
- » It produces cluster of orange coloured tubular flowers, produces dark red coloured berries.
- » The birds are attracted by the fruits and act as dissemination agents.
- » The seeds dropped on the crevices of the stem germinate and produce haustoria to attach to the host and proliferate in large numbers.

Parasite: *Dendrophthoe falcata*

Management

- » Removal of parasite affected branches and apply Bordeaux paste at cut ends.
- » Collect the seeds of parasite and destroy.

Other Diseases

4. **Canker and Dieback:** *Botryosphaeria parva, Botryosphaeria dothidea, Botryosphaeria quercuum*
5. **Leaf spot**: *Amphichaeta grevilleae*
6. **Die-back**: *Erythrecium salmonicolor, Phyllosticta telopeae*

Selected References

Johnson JD, Appel DN. 2006. Major Oak Diseases and their Control, Garden Fact Sheet, Texas Agricultural Extension Service, Texas AandM University.

Summerell BA. 2017. Diseases of Proteaceae. In: McGovern R., Elmer W. (eds) Handbook of Florists' Crops Diseases. Handbook of Plant Disease Management. Springer, Cham

Venkataramani KS. 1954. A leaf disease of nursery plants of the Silver Oak. Planters' Chronicle, Coonoor 49, 570-571.

28 Diseases of *Hardwickia binata* (Indian blackwood)

Wilt

Symptoms

- The disease symptoms started at tip of the plant and along the margin of the leaves.
- Defoliation and eventually death of the plants.

Fungus: *Fusarium oxysporum*

Management

- All dead, dry stumps and drying trees should be removed.
- Spot drenching with Carbendazim 0.1% or Metalaxyl + Mancozeb 0.2% as twice at 15 days interval.

Selected References

Rai VR, Mamatha T. 2005. Seedling diseases of some important forest tree species and their management. Working Papers of the Finnish Forest Research Institute 11

29 Diseases of *Jatropha curcas* (Physic nut)

S. No.	Disease	Pathogen
1.	Damping off	*Fusarium solani*
2.	Root rot	*Macrophomina phaseolina*
3.	Powdery mildew	*Oidium* sp.
4.	Leaf blight	*Phoma* sp.
5.	Anthracnose	*Colletotrichum gloeosporioides*
6.	*Pestalotiopsis* stem canker	*Pestalotiopsis mangiferae*
7.	Leaf spot	*Alternaria ricini*
8.	Rust	*Melampsora ricini, Phakopsora jatrophicola*
9.	*Cercospora* leaf spot	*Cercospora jatrophicola*
10.	*Fusarium* wilt	*Fusarium oxysporum*
11.	*Botrytis* rot	*Botrytis ricini*
12.	Bacterial spot	*Xanthomonas ricinicola*

1. Damping off

Symptoms

» The disease starts as water soaked black lesion at the collar region of the young seedling, which later rot and topple over the soil.

» Light pink coloured mycelial masses can be seen on the soil surface.

Fungus: *Fusarium solani*

Management

» Soil drenching with Carbendazim 0.1%

2. Root rot

Symptoms

» The disease starts as water soaked black lesion at the collar region of the plant near soil level and progresses both upward and downward.

» The lower leaves turn yellow and wither away.

» This is followed by sudden dead of the whole plant.

» The dried plant show extensive rooting of the root system with disintegrated barks.

» The close observation of the shredded bark tissues shows large number of black colored resting structures of the fungus.

» The stem portion above soil level becomes ashy brown indicating the presence of thousands of minute black dot like fungal fructification.

Fungus: *Macrophomina phaseolina*

Management

» Soil drenching with Carbendazim 0.1%.

» Soil drenching with *Trichoderma viride* at 2.5 kg/ 50 kg FYM and apply 200 g /plant.

3. Powdery mildew

Symptoms

» The powdery mildew disease develops as a minute white patch on the upper surface or lower surface, which later extends to cover the entire leaf lamina.

» As the disease progresses the powdery growth turns to light brown patches leading to defoliation.

- The infection may also occur on the flower stalk and flowers resulting in the shedding of flowers and young fruits.
- The white powdery growth of the pathogen represents the mass of conidiophore and conidia.
- In the advanced stage of the disease the fungi produce the sexual fruiting body called Cleitothecia.

Fungus: *Oidium* sp.

Management

- Spray with Wettable sulphur 2 g/ ml or Chlorothalonil 0.1% or Calixin 0.1%.

4. Leaf blight

Symptoms

- The symptoms appear as minute dark brown dots, 2-3mm in diameter.
- Spots enlarge to 5 to 8 mm in diameter and turn to light pale brown with dark brown outline.
- The margin grow out wards forming one to three dark brown concentric rings around the light coloured centre spot with a lot of black pycnidia.

Fungus: *Phoma* sp.

Management

- Foliar spray with Copper oxychloride 0.5 g / lit or Mancozeb 0.2% twice at 15 days interval.

5. Anthracnose

Symptoms

- The disease initiates as a minute black spot on the leaf surface and enlarges to form a spot with a black margin showing a large number of acervuli in the centre.
- The black lesions also appear on the young shoots showing die back symptoms.
- The infection on the flower stalk leads to flower shedding. Black coloured spots also appear on the surface of the fruit showing fruit rot symptoms.

Fungus: *Colletotrichum gloeosporioides*

Management

» Foliar spray with Copper oxychloride 0.2% or Mancozeb 0.2% twice at 15 days interval.

6. *Pestalotiopsis* stem canker

Symptoms

» Stems of affected plants had multiple injuries at the collar region and above.

» The exposed inner tissue was dark brown to black.

» At these wounds, small to extended cankers were formed giving the area a charcoal black appearance.

» Minute black dots were also visible on the light coloured exposed tissue.

Fungus: *Pestalotiopsis mangiferae*

Management

» Foliar spray with Carbendazim 0.1% or Mancozeb 0.3% twice at 15 days interval.

7. Leaf spot

Symptoms

» Initially the attack appears on the leaves.

» If the humidity is high, this disease develops very quickly on the fruit capsule and the fruit become black.

» If the attack occurs in the beginning of flower forming, the buds may die.

» And if it occurs at the end of flowering, the flower is opened but no fruit capsule is formed.

Fungus: *Alternaria ricini*

Management

» Foliar spray with Mancozeb 0.3% twice at 15 days interval.

8. Rust

Symptoms

- The disease looks like rust spots on the lower surface of the leaf and yellow in color
- A heavy damage may dry the leaf

Fungi: *Melampsora ricini, Phakopsora jatrophicola*

Management

- Foliar spray with Oxycarboxin 0.1% twice at 15 days interval.

9. Cercospora leaf spot

Symptoms

- The damage caused by this disease on *Jatropha curcas* can be very heavy.
- The symptom is that black spot or brown spot surrounded by ring with pale green in color.
- The spots are found on both leaf surfaces.
- When the spot is getting bigger, the spot is become gray, surrounded by brown color.
- Initially the spot is round and finally the spot form is irregular.

Fungi: *Cercospora jatrophicola*, *Cercospora jatrophigena*, *Pseudocercospora jatrophae-curcas*, *Pseudocercospora jatrophae*, *Pseudocercospora jatropharum*.

Management

- Foliar spray with Carbendazim 0.1% or Mancozeb 0.3% twice at 15 days interval.

10. Fusarium wilt

Symptoms

- If the seedling is attacked, the leaves become pale green, and withered and finally died.
- Leaves on the lower part of plant will fall and only leaves on the upper part will leave.
- Complete death of the plant was observed.

Fungus: *Fusarium oxysporum*

Management

» Foliar spray with Mancozeb 0.2% + Thiophanate methyl 0.1% twice at 15 days interval.

***11. Botrytis* rot**

Symptoms

» Initially the symptom is blackish spot on the flower

» It becomes a serious problem in the rainy season when the fruit capsule is formed

» The disease is well developed in a cool temperature and rain

» The infected flower will be spoiled and covered by gray mold.

Fungus: *Botrytis ricini*

Management

» Foliar spray with Trifloxystrobin 0.1% or Tebuconazole 0.1% twice at 15 days interval.

12. Bacterial spot

Symptoms

» This disease is caused by bacteria which may attack the cotyledon and the leaf

» The symptom is the black spots, round and irregular in form

Bacterium: *Xanthomonas ricinicola*

Management

» Foliar spray with Streptocyclin sulphate.

Other Diseases

13. Anthracnose: *Colletotrichum gloeosporioides, Colletotrichum capsici*

14. Dieback / Stem canker: *Lasiodiplodia theobromae*

15. **Powdery mildew**: *Pseudoidium jatrophae*

16. ***Fusarium* wilt**: *Fusarium oxysporum*

17. **Collar rot / root rot**: *Macrophomina phaseolina*

18. ***Botrytis* rot**: *Botrytis ricini*

19. **Dieback**: *Fusarium* sp.

20. ***Pestalotiopsis* leaf blight**: *Pestalotiopsis versicolor*

21. ***Verticillum* leaf blight**: *Verticillum dahliae*

22. ***Cylindrocladium* leaf blight/ White mould**: *Cylindrocladium jatrophae*

23. *Jatropha mosaic India virus*

Selected References

Anitha K, Varaprasad KS. 2012. Jatropha Pests and Diseases: An Overview. Jatropha, Challenges for a New Energy Crop, 175–218.

Bock KR, Guthrie EJ, Figueiredo G 1981. A strain of cassava latent virus occurring in coastal districts of Kenya. Ann Appl Biol 99:151–159

Bocourt PY, López Manes D, López Mesa MO. 2005. New host plants for the family Erysiphaceae in Cuba. Fitosanidad 9:73

Braun U, Freire F, Das CO 2004. Some cercosporoid hyphomycetes from Brazil—III. Cryptogam Mycol 25:221–244

Braun U. 1987. A monograph of the Erysiphales (powdery mildews). Beih Nova Hedwigia 89:700

Brown JK, Idris AM, Torres JI, Bird J. 1999. Jatropha mosaic begomovirus variants from weed and cultivated hosts in Puerto Rico. Phytopathology 90:S122

CABI UK Centre Annual Report. 2009. Assessing the impact of diseases on the biofuel crop Jatropha.

Carels N. 2009. *Jatropha curcas*: a review. In: Delseny M, Kader JC (eds) Advances in botanical research. Elsevier, Amsterdam, pp 39–86 Chattopadhyay NC (2009) Disease and pest of Jatropha curcas in West Bengal, India.

Chavhan TL, Suryanarayana V, Naik ST. 2010. Survey and management of Pestalotiopsis leaf blight of Jatropha a destructive new disease in Karnataka. Indian Phytopathol 63:110–111.

Das AK, Chattopadhyay BK. 1990. Three new combinations into the genus *Pseudocercospora* Speg. J Mycopathol Res 28:27–32.

Das N. 1995. *Jatropha glandulifera* Roxb.—a new host for *Fusarium solani* (Martius) Sacc. Indian J Mycol Plant Pathol 25:337

Gao SQ, Qu J, Chua NH, Ye J. 2010. A new strain of Indian cassava mosaic virus causes a mosaic disease in the biodiesel crop *Jatropha curcas*. Arch Virol 155:607–612

Garga RP. 1960. A leaf distorting virus disease of *Jatropha curcas* Linn. Curr Sci 30:345–346.

Gupta PK, Sharma ND, Singh SR, Singh OP. 2007. Fungi associated with medicinal plants of Madhya Pradesh. Ann Plant Prot Sci 15:508–509

Hosagoudar VB, Archana GR. 2009. Host range of meliolaceous fungi in India. J Threat Taxa 1:269–282

Hudge BV, Datar VV. 2009) In vitro effect of temperature and humidity against Alternaria alternata (Fr.) Keissler causing leaf spot of Jatropha. Ann Plant Physiol 23:129–130

Hudge BV, Datar VV. 2010. Study of incidence and severity of leaf spot disease in Jatropha curcas L. Intl J Agric Sci 6:355–356

Jones MH, Csurhes S. 2008. Pest Plant Risk Assessment: Physic Nut Jatropha curcas, Department of Primary Industries and Fisheries, the State of Queensland: 1–28

Kar AK, Das A. 1988. New records of fungi from India. Indian Phytopathol 41:505.

Kirk PM. 1991. *Nigrospora sphaerica* [Descriptions of Fungi and Bacteria]. IMI descriptions of fungi and bacteria no. 106 pp. Sheet 1056. UK, CAB International

Kumar S, Sharma S. 2010. Studies on the factors affecting pathogenicity of root rot caused by *Rhizoctonia bataticola* in *Jatropha curcas*. Indian For 136:736–741

Latha P, Prakasam V, Kamalakannan A, Gopalakrishnan C, Raguchander T, Paramathma M. 2009. First report of *Lasiodiplodia theobromae* (Pat.) Griffon and Maubl. causing root and collar rot disease of physic nut (*Jatropha curcas* L.) in India. Australasian Plant Dis Notes 4:19–21

Li Q, Zhao YB, Gao TP, Zhao GY, Xu H. 2009. Preliminary Report on Disease and Insect Pests of Jatropha carcass Nursery. For Inventory and Plann 01–019.

Machado AR, Pereira OL. 2012. Major Diseases of the Biofuel Plant, Physic Nut (Jatropha curcas), Biodiesel - Feedstocks, Production and Applications, Zhen Fang, IntechOpen, DOI: 10.5772/52336.

Pereira OL, Dutra DC, Dias LAS. 2009. *Lasiodiplodia theobromae* is the causal agent of a damaging root and collar rot disease on the biofuel plant Jatropha curcas in Brazil. Australasian Plant Dis Notes 4:120–123

Phillips S. 1975. A new record of *Pestalotiopsis versicolor* on the leaves of *Jatropha curcas*. Indian Phytopathol 28:546.

Raj SK, Kumar S, Snehi SK, Pathre U. 2008. First report of Cucumber mosaic virus on Jatropha curcas in India. Plant Dis 92:171.

Reddy JR, Reddy AP. 1980. *Erysiphe euphorbiae* sp. on *Jatropha gossypiifolia* L. Curr Microbiol 4:95–97.

Seier M, Cortat G, Hill L. 2009. Preliminary assessment of the rust, *Phakospora jatrophicola*, as a potential biocontrol agent for *Jatropha gossypiifolia*. p 7 in annual report, 2009.

Sharma PK, Sangle UR, Ahmad N. 2010. A new record of Oidium state of *Erysiphe jatrophe* Doidge from Manipur causing powdery mildew on Jatropha. J Mycol Plant Pathol 40:467

Sharma S, Kaushik JC, Kaushik N. 2001. *Fusarium moniliforme* causing root rot of jatropha. Indian Phytopathol 54:275

Shukla R, Jamaluddin. 2010. Two new leaf spot diseases of *Jatropha curcas* L from central India. J Mycol Plant Pathol 40:479–480.

Sidhu OP, Sanjay A, Uday P, Snehi SK, Raj SK, Roy Raja. 2010. Metabolic and histopathological alterations of Jatropha mosaic begomovirus -infected Jatropha curcas L. by HR-MAS NMR spectroscopy and magnetic resonance imaging. Planta 232:85–93

Singh ID. 1983. New leaf spot diseases of two medicinal plants. Madras Agric J 70:490 Smith KM. 1957. A text book of plant virus diseases, 2nd edn. Churchill, London

Zeigler RS, Lozano JC. 1983. The relationship of some Sphaceloma species pathogenic on cassava and other Euphorbiaceae in Central and South America. Phytopathol 73:293–300.

30 Diseases of *Leucaena leucocephala* (River Tamarind)

S. No.	Disease	Pathogen
1.	Leaf spot	*Cercospora cruenta*
2.	Leaf blight	*Alternaria alternata*
3.	Sooty mould	*Capnodium indicum*
4.	Root rot	*Ganoderma lucidum*

1. Leaf spot

Symptoms

- Small circular spot with a re margin develops on old leaves.
- As the disease advances, the centre of spot become white, thin and drops off leaving shot holes.

Fungus: *Cercospora cruenta*

Management

- Removal of infected plant debris.
- Spraying of Carbendazim 0.1% or Copper oxy chloride 0.05% or Chlorothalonil 0.1% or Mancozeb 0.1%.

2. Leaf blight

Symptoms

- » A brown water soaked lesion develops on the leaf and spreads fast to occupy major portion of the leaf leading to complete drying of the lamina.
- » The disease can easily identified by presence of concentric rings in the blighted area showing the sporulation of the fungus.
- » The fungus produces brown coloured conidia with both vertical and horizontal septa with a beak at the top.

Fungus: *Alternaria alternata*

Management

- » Removal of infected plant debris.
- » Prophylactic sprays with Carbendazim + Mancozeb 0.2%.

3. Sooty mould

Symptoms

- » The infection appears on the upper surface of the leaf as a sparse, black network of hyphae.
- » Then a thin effuse black powdery coating can be seen covering a major portion of the leaf.
- » This indicates the sporulation of the fungus.
- » The heavy infestation of the sucking pests leads to development of the sooty growth on the leaf surface.

Fungus: *Capnodium indicum*

Management

- » Spray starch solution 5%.

4. Root rot

Symptoms

- Pathogen enters through distal root end.
- Pale leaves which later dries, white mycelial strands seen in between bark and wood.
- When most of roots are affected, trees are killed.

Fungus: *Ganoderma lucidum*

Management

- Maintain the trees in healthy condition – application of manure.
- Do not wound the tree. If any wounds are there apply Bordeaux paste.
- Remove the sporocarp and destroy.
- Scrap the mycelia portion and apply Bordeaux paste.
- Remove severely affected trees.
- Isolate the affected tree from healthy trees.
- Drench the soil near the tree with 40 litres of 1% Bordeaux mixture at quarterly interval for thrice a year and repeat after 2-3 years.
- Root injection in the infected plant containing 2g of Aureofungin soil and 1 g of copper sulphate in 100 ml of water.

Other Diseases

5. **Pod rot**: *Fusarium* sp.
6. **Leaf spot**: *Camptomeris leucaenae, Cercospora cruenta*
7. **Leaf blight**: *Alternaria alternata*
8. **Sooty mould**: *Capnodium indicum*
9. **Stem canker**: *Pirex subrinosus*

Selected References

Lenné JM. 1991. Diseases of Leucena species. Tropical Pest Management 37, 281–9.

Sridhar KR. 2008. Frontiers in Fungal Ecology, Diversity and Metabolites, IK International Publishing House Pvt. Ltd.

Thiribhuvanamala G. 2017. Major diseases affecting pulp wood trees and their management. In: K.T. Parthiban, R. Seenivasan (Eds) Plantation and Agroforestry Pulpwood Value Chain Approach, Scientific Publisher, India

31 Diseases of *Madhuca longifolia* var. *latifolia* (Mahua)

S. No.	Disease	Pathogen
1.	Grey leaf spot	*Pestatiopsis mathucae*
2.	Leaf blight	*Phoma* sp.
3.	Sooty mould	*Capnodium indicum*
4.	Rust	*Scopella echinulata*
5.	Red rust	*Cephaleuros parasiticus*

1. Grey leaf spot

Symptoms

- Grey coloured irregular lesion normally develops near the tip of the leaf or margin, which advances to occupy major portion of the leaf area showing blighted appearance.
- The necrotic dried spots large number of black dots can be seen, they are the fruiting body of the fungus namely acervulus.

Fungus: *Pestatiopsis mathucae*

Management

- Spray with Carbendazim + Mancozeb 0.2% twice at 15 days interval.

2. Leaf blight

Symptoms

- Spots appear as minute dark brown dots and later enlarge and turn to light pale brown with dark brown outline.
- The margin grows out wards forming one to three dark brown concentric rings around the light coloured centre spot.

Fungus: *Phoma* **sp.**

Management

- Spray with Copper oxychloride 0.5 g / lit or Mancozeb 0.2% twice at 15 days interval.

3. Sooty mould

Symptoms

- The infection appears on the upper surface of the leaf as a sparse, black network of hyphae.
- Then a thin effuse black powdery coating can be seen covering a major portion of the leaf.
- This indicates the sporulation of the fungus.
- The heavy infestation of the sucking pests leads to development of the sooty growth on the leaf surface.

Fungus: *Capnodium indicum*

Management

- Spray Imidacloprid 0.5% in nursery to manage sucking pests.
- Spray starch solution 5%.

4. Rust

Symptoms

- Raised rusty pustules observed on the lower surface of the leaf.

- Rust has dark brown telutospores on lower surface of leaf.
- Telutospores were observed are spiny and irregular round in shape.
- The rust formed dark chocolate color pustules on lower surface of the leaf.

Fungus: *Scopella echinulata*

Management

- Foliar spraying of Carboxin 0.1% or Triadimenol 0.2%.

5. Red rust or Algal spot

Symptoms

- Reddish circular spots with slightly radiating margin develop on the upper surface of the leaves, the affected spots turn to greyish green in colour and leaf shedding occurs in severe infection.

Alga: *Cephaleuros parasiticus*

Management

- Foliar spraying of Bordeaux mixture 1%.

Selected References

Pathak H, Maru S, Satya HN and Silawat SC. 2015. Fungal Diseases of Trees in Forest Nurseries of Indore, India. Journal of Plant Pathology and Microbiology. 6:8 DOI: 10.4172/2157-7471.1000297

32 Diseases of *Magnolia champaca* (Champak)

1. Leaf spot

Symptoms

- » The occurrence of a leaf spot disease late in the growing season generally does not seriously affect the health of a tree.
- » The diseases develop as small, scattered, circular to oval dead areas in the leaves; usually tan, dark brown, yellow, gray, purple, or black.
- » Heavily infected leaves may turn yellow to brown, wither, and drop early, weakening the tree.
- » Occasionally, some leaf spotting fungi deform or kill flowers, buds, fruits, twigs, or even small branches.

Fungus: *Phomopsis micheliae*

Management

- » Foliar spraying of carbendazim 0.1% or copper oxy chloride 0.25% or chlorothalonil 0.1% or mancozeb 0.2%.

Selected References

Sridhar KR. 2008. Frontiers in Fungal Ecology, Diversity and Metabolites, IK International Publishing House Pvt. Ltd.

Sankaran KV, Florence EJM, Sharma JK. 1987 – Two new species of *Phomopsis* from India. Transactions of the British Mycological Society 89, 404–407

33 Diseases of *Melia dubia* (Malabar neem)

S. No.	Disease	Pathogen
1.	Collar rot and seedling web blight	*Rhizoctonia solani*, *Pythium* sp.
2.	Leaf spot	*Colletotrichum dematium*, *Cylindrocladium ilicicola*
3.	Root rot	*Lasiodiplodia theobromae*
4.	Wilt	*Fusarium moniliforme*

1. Collar rot and seedling web blight

Symptoms

» The disease appeared as water-soaked lesions on seedling stem at collar region, which spread longitudinally and became necrotic in due course.

» The infection also affects the apical portion of the seedlings, *viz.*, stem and foliage and showed the typical symptoms of web blight.

» Water-soaked lesions appeared on lower leaves, which coalesced and spread to the entire leaf lamina.

» The affected leaves became necrotic and caused premature defoliation.

Fungus: *Rhizoctonia solani*

Oomycete: *Pythium* sp.

Management

- Soil drenching with *Trichoderma viride* along with FYM or neem cake.
- Soil drenching with Carboxin 0.2%.

2. Leaf spot

Symptoms

- The symptoms produced by both the pathogens are almost similar; disease appeared as small pin prick lesions, pale brown in colour, later coalesced and spread to form circular necrotic lesions.
- In severe case, the infection spreads to the entire leaf lamina.
- Premature defoliation was noticed in container seedlings.

Fungi: *Colletotrichum dematium*, *Cylindrocladium ilicicola*

Management

- Spray with Carbendazim 0.2% at fortnight interval.

3. Root rot

Symptoms

- Suddenly plant becomes dry and dead.
- Dark black lesions extend at the collar region (joining the stem and root).
- Later bark shredding is observed in the root portion.
- Branches and stems may be partially or completely girdled, causing crown dieback andpossibly tree death.
- Minute black pycnidia are seen on the affected portions.

Fungus: *Lasiodiplodia theobromae*

Management

- Provide proper spacing and thinning.

- Drenching with Carbendazim 0.2%.
- Application of *Trichoderma viride* 200 g/ plant.

4. Wilt

Symptoms

- In nursery, plants exhibit yellowing and withering symptoms.
- Affected seedlings die rapidly in mass mortality.

Fungus: *Fusarium moniliforme*

Management

- Provide proper spacing and thinning.
- Drenching with Copper oxy chloride 0.25% or Bordeaux mixture 1%.
- Application of *Trichoderma viride* 200 g/ sq feet.

Selected References

- Warrior R. 2011. *Melia dubia,* ICFRE-IFGTB.
- Muthulakshmi P, Naresh S, Rajamanickam S, Parthasarathy S. 2016. Identification and morphological characterization of pathogens infecting *Melia dubia* in Tamil Nadu. International Journal of Research in Applied, Natural and Social Sciences, 4(11):191-198.
- Muthulakshmi P, Parthasarathy S. 2019. Evaluation of Fungicides and Bio-Control Agents against Foliar and Soil Borne Diseases of *Melia dubia*. International Journal of Current Microbiology and Applied Sciences, 8(9): 2476- 2486.

34 Diseases of *Millettia pinnata* (Indian beech)

S. No.	Disease	Pathogen
1.	*Cercospora* leaf spot	*Cercospora pongamiae*
2.	*Pseudocercospora* leaf spot	*Pseudocercospora pongamiae-pinnatae*
3.	Leaf blight	*Alternaria alternata*
4.	Leaf blight	*Phoma indica*
5.	Powdery mildew	*Oidium* sp.
6.	Leaf rust	*Ravenelia hobsoni*
7.	Sooty mould	*Capnodium braziliense*
8.	*Fusicladium* leaf spot and blight	*Fusicladium pongamiae*
9.	*Colletotrichum* leaf spot and blight	*Colletotrichum* sp.
10.	*Dothiorella* leaf blight	*Dothiorella* sp.
11.	*Rhytisma* tar spot	*Rhytisma* sp.

1. Cercospora leaf spot

Symptoms

» Small circular spot with a re margin develops on old leaves.

» As the disease advances, the centre of spot become white, thin and drops off leaving shot holes.

» In severe infection defoliation also occurs.

Fungus: *Cercospora pongamiae*

Management

» Spray with Mancozeb 0.2% or Copper oxychloride 1.5 g/ lit twice at 15 days interval.

2. *Pseudocercospora* leaf spot

Symptoms

» Lesions subcircular to angular, mostly small, up to 1 mm wide, discrete, sometimes coalescing to form spots 5 mm diam., pale brown with dark brown margins usually surrounded by a yellow halo.

Fungus: *Pseudocercospora pongamiae-pinnatae*

Management

» Spray with Mancozeb 0.2% or Carbendazim 0.1% twice at 15 days interval.

3. Leaf blight

Symptoms

» A brown water soaked lesion develops on the leaf and spreads fast to occupy major portion of the leaf leading to complete drying of the lamina.

» The disease can easily identified by presence of concentric rings in the blighted area showing the sporulation of the fungus.

Fungus: *Alternaria alternata*

Management

» Spray with Carbendazim + Mancozeb 0.2% twice at 15 days interval.

4. Leaf blight

Symptoms

» Spots appear as minute dark brown dots and later enlarge and turn to light grey centre with dark brown outline.

» Minute pycnidia are seen in the centre of the spot.

Fungus: *Phoma indica*

Management

» Spray with Mancozeb 0.2%.

5. Powdery mildew

Symptoms

» The pathogen formed irregular white patches, consisting of mycelium and asexual conidia on the surface of the leaves.

» These patches coalesced and covered the whole leaf lamina giving grayish white appearance.

» Severely infected leaves and leaflets defoliated prematurely.

Fungus: *Oidium* sp.

Management

» Spray of Carbendazim 0.2% or Chlorothalonil 0.2% found to be effective in minimizing the disease.

6. Leaf rust

Symptoms

» The rust fungus infects the leaves and produces initially orange yellow pustules which later shows numerous chestnut brown teliospore heads on the lower surface of the leaves.

Fungus: *Ravenelia hobsoni*

Management

» Dusting or foliar spray of sulphur based fungicide 0.05%.

7. Sooty mould

Symptoms

» The infection appears on the upper surface of the leaf as a sparse, black network of hyphae.

- Then a thin effuse black powdery coating can be seen covering a major portion of the leaf.
- This indicates the sporulation of the fungus.
- The heavy infestation of the sucking pests leads to development of the sooty growth on the leaf surface.
- Sucking pests secrete sugary solution that facilitates sooty mould growth.

Fungus: *Capnodium braziliense*

Management

- Spray with Imidacloprid 3 g / lit in nursery to manage sucking pests.
- Spray starch solution 5%.

8. Fusicladium leaf spot and blight

Symptoms

- Chlorotic translucent irregular areas dorsally on leaves followed by development of distinct dot sized black eruptive clusters on ventral surface with corresponding pale green or greenish yellow areas on dorsal side.
- Black coloured blighting, inward curling and premature defoliation was ultimate.

Fungus: *Fusicladium* sp.

Management

- Spray with Difenconazole 0.1% or Carbendazim 0.2%.

9. Colletotrichum leaf spot and blight

Symptoms

- Brown spot with yellow halo from margin towards centre of leaf lamina of matured leaves followed by distinct brown coloured blighting and inward rolling.
- Premature defoliation was distinct in final stage.

Fungus: *Colletotrichum* sp.

Management

» Spray with Propiconazole 0.1% or Hexaconazole 0.1% or Difenconazole 0.1% or Carbendazim 0.2%.

10. Dothiorella leaf blight

Symptoms

» Papery textured creamy white blighted areas possessing black dot like structures along leaf margins followed by blighting of entire leaf and its defoliation.

Fungus: *Dothiorella* sp.

Management

» Spray with Difenconazole 0.1% or Carbendazim 0.2%.

11. Rhytisma tar spot

Symptoms

» Black tar like spots with distinct orange halo followed by premature yellowing, drying and defoliation.

Fungus: *Rhytisma* sp.

Management

» Spray with Propiconazole 0.1% or Hexaconazole 0.1% or Carbendazim 0.2%.

Selected References

Arpiwi NL, Yan G, Barbour EL. 2011. Genetic diversity, seed traits and salinity tolerance of *Millettia pinnata* (L.) Panigraphi (syn. *Pongamia pinnata*), a biodiesel tree. in review. J. Tropical For. 23, 51-55.

Borah RK, Dutta D, Hazarika P. Some new records of fungi from North East India. Bano Biggyan Potrika. 1998; 27:41-13.

Jamaluddin, Goswami MS, Ojha BM. 2004. Fungi of India 1989-2001. Scientific Publisher, Jodhpur, India

Meeta M and Jindal KK. 1994. Diseases of Ornamental Plants in India, Daya Publishing House

RaghuRam M. Mallaiah KV. 1993. *Pseudocercospora pongamiae-pinnatae* sp. nov. from India, Mycol. Res. 97 (1), 127-123

Shivanna MB. 2005. Fungal diseases in forest nurseries in Shimoga district, Karnataka, India. Working Papers of the Finnish Forest Research Institute 11

35 Diseases of *Morus indica* (Mulberry)

S. No.	Disease	Pathogen
1.	Powdery mildew	*Phyllactinia corylea, Phyllactinia guttata*
2.	Leaf rust	*Peridiospora mori*
3.	*Cercospora* leaf spot	*Cercospora moricola, Cercospora moricola, Cercospora missouriensis, Cercospora snelliana*
4.	*Pseudocercospora* leaf spot	*Pseudocercospora mori*
5.	Anthracnose	*Colletotrichum demantium*
6.	*Fusarium* root rot	*Fusarium oxysporum*
7.	Sooty mould	*Capnodium* sp.
8.	Bacterial leaf spot	*Pseudomonas syringae* pv. *mori*
9.	Mosaic	*Mulberry mosaic virus*

1. Powdery mildew

Symptoms

» White powdery patches appear on the lower surface of leaf which is gradually increased and cover whole leaf surface.

» Affected leaves turn yellowish to purplish brown and defoliate prematurely.

Fungi: *Phyllactinia corylea, Phyllactinia guttata*

Management

- » Grow resistant varieties like Mandania, Katania, China-White, Jodhpur, Punjab local, MR-1, MR-2, S 31, S 54, S 153, S 796, S 1096, Kanva-2, Calbresa, Cattaneo, Shrim-2, ACC123, ACC-125, ACC-151, ACC153, OPH-3, Almora local, Himachal local and 5-523.
- » Provide wider spacing 3' × 3' between rows and plants, keeping the single trunk of height 30 cm.
- » Spray with Bordeaux mixture 1% or Carbendazim 0.1% or Morestan 0.025%.

2. Leaf rust

a) *Peridiospora mori (*Syn: *Aecidium mori)*

Symptoms

- » Several small pin head shaped, reddish brown, irregular pustules appear on the both surface of mature leaves and green woody portions of stem.
- » Severely infected leaves turn yellowish, margin of the leaves become dry and fall prematurely.

Fungus: *Peridiospora mori*

b) *Cerotelium fici*

Symptoms

- » Raised rusty pustules appeared on the lower surface of the leaves.
- » In severe, leaves are deformed and fall off.

Fungus: *Cerotelium fici*

Management

- » Grow resistant varieties like China Peking, Cattaneo and BC 259.
- » Provide wider spacing 3' × 3' between rows and plants.
- » Spray with Captafol 0.2% or Chlorothalanil 0.2% or Mancozeb 0.1% or Propineb 0.2%.

3. *Cercospora* leaf spot

Symptoms

- Minute, circular or irregular light brown spots appear on both sides of the leaves.
- The adjacent spots unite together to form a larger spot.
- The necrotic tissues of such spots drop out and form the characteristics shot holes.
- Highly infected leaves become yellowing and defoliate prematurely.

Fungi: *Cercospora moricola, Cercospora moricola, Cercospora missouriensis, Cercospora snelliana*

Management

- Grow resistant varieties like Kalakuthai, Bidevalaya, Kanva-2, S-54, C-799, Shrim-2, Moulai, Mizusawa, Assamabola, Cattaneo, Miruso, ACC-106, ACC-109, ACC-122, ACC-114, ACC-115, ACC-116, ACC-117, ACC-119, ACC-121, ACC-123, ACC-125, ACC-128, ACC-152, ACC-153, ACC-201, OPH-1, OPH-3, MS-2, MS-3, MS-5, MS-7, MS-8, Paraguay, RFS-135, RFS-175, S-153, Almora Local and S-1096.
- Provide wider spacing.
- Spray with Carbendazim 0.025%.

4. *Pseudocercospora* leaf spot

Symptoms

- Small grey to brown colored hyphal mats on lower surface of the leaves, and deep brown colored spots develop on the upper surface.
- Lower most leaves exhibit yellowing and defoliation.
- The necrotic tissues of such spots drop out and form the characteristics shot holes.
- Highly infected leaves become yellowing and defoliate prematurely.

Fungus: *Pseudocercospora mori*

Management

- Grow resistant varieties like C-763, C-1726, C-1635, Tr-4 and Tr-10.
- Spray with Carbendazim 0.1%.

5. Anthracnose

a) *Colletotrichum demantium* (Syn: *C. morifolium*)

Symptoms

» Initially faint brown lesions on leaf, which enlarge and transform into oval, angular or irregular shaped spots with black brown borders.

» It produces marginal scorching, drying, yellowing and fall prematurely.

Fungus: *Colletotrichum demantium*

b) *Colletotrichum gloeosporioides* (Teleomorph: *Glomerella cingulata*)

Symptoms

» This is also known as black leaf spot disease.

» Initially coral red to brown color spots appear on leaf blade, mostly towards the margin and along the veins.

» In severe, leaves are deformed and fall off.

Fungus: *Colletotrichum gloeosporioides*

b) *Colletotrichum lindemuthianum*

Symptoms

» Initially brown to reddish brown, round or oval and depressed eyespots on the leaf blade, spread towards the margin.

Fungus: *Colletotrichum lindemuthianum*

Management

» Grow resistant varieties like Kosen, S 146 and BC 259.

» Grow black leaf spot resistant varieties like China Peking, MR 2, ACC 151 and ACC 233.

» Spray with *Bacillus amyloliquifaciens* commercial formulation.

» Spray with Carbendazim 0.025% or Thiram 0.1% or Mancozeb 0.2% or Propineb 0.2%.

6. *Fusarium* Root Rot

Symptoms

- The leaves are suddenly withering and defoliation.
- The bark of the infected root is decayed and easily of peel off.
- The surface of root and cortex appears gummy and the plant dies.

Fungi: *Fusarium oxysporum*, *Fusarium solani*

Management

- Diseased plants should be uprooted and burnt.
- The field should be flooded to destroy dormant mycelia and spores.
- Land reclaimed by placing chloropicrin or calcium cyanamide 2.5 kg / m^2.
- Soil disinfection by di-trapex or vorlex (80% chlorinated C3 hydrocarbon + 20% methyl isothiocyanide).
- Planting pits can be inoculated with bioagent (*Trichoderma harzianum* 1 kg + FYM 50 kg) at 500 g/ pit.
- Planting pits can be dusted with mancozeb 2%.
- Seedlings dip with carbendazim 0.2%.
- Provide wider spacing 3' × 3' between rows and plants.
- Spray with Captafol 0.2% or Chlorothalanil 0.2% or Mancozeb 0.1% or Propineb 0.2%.

7. Sooty mould

Symptoms

- Thick black coating developed on the upper surface of the leaves.
- Associated with mulberry white fly (*Bemesia myricae*)

Fungi: *Capnodium* sp., *Chaetothyrium* sp., *Curvularia affinis*

Management

- Spray with Imidacloprid 3 g / lit in nursery to manage sucking pests.
- Spray starch solution 5%.

8. Bacterial blight

Symptoms

a. Shrunk leaf type or Halo blight

- Green water soaked irregular spots appeared on the leaves, bacterial ooze out on the lower surface.
- Later the spots become brown, blighted, irregular or slightly angular and surrounded by fade green or yellow halo.
- In severe, leaves are wrinkled, distorted, shrunk backwards and eventually defoliate.
- On the bark, appears as black, sunken and shuttle like lesions.
- It is common during rainy season when there is high humidity and temperature.

b. Blackish bacterial blight or Head rot or Plague.

- It appears as semi-transparant spots on young shoots, later turn to blackish brown in color.
- The xylem becomes brown and rotten.

Bacterium: *Pseudomonas syringae* pv. *mori*

Management

- Grow resistant varieties like Kosen, S-36, S-41 and S-54.
- Spray with Captafol 0.2% or Mancozeb 0.2%.

9. Mosaic

a) *Mulberry mosaic virus*

- Infected leaves shows typical mosaic mottling with slight curling and puckering.

Vector: *Aphis gossypii* Glover, *Myzus persicae* and *Rhopalosiphum maidis* (Fitch) (Aphids).

b) *Fig mosaic virus*

- Infected leaves shows faint chlorotic spots followed by systematic mottling.

Vector: *Aceria ficus* Cotte (Eriophyid mite).

Management

- Grow resistant varieties like Ichinose, Kairyonezumegaishi of *M. alba* and Oshimaso and Kosen on *M. latifolia*.
- Spray with Carbendazim 0.1%.

Minor Diseases

10. **Fungal leaf blight:** *Alternaria alternata, Bipolaris tetramera, Fusarium pallidoroseum*
11. **Leaf blotch:** *Srirosporium mori*
12. **Leaf spot:** *Alternaria tenuissima, Phleospora maculans, Coniothyrium albae, Dreshlera yamadi, Fusarium concolar, Fusarium solani, Phoma mororum, Phyllosticta morifolia.*
13. **Tar leaf spot:** *Myrothecium roridum*
14. **Charcoal rot:** *Macrophomina phaseolina*
15. **Collar rot:** *Phoma mororum, Sclerotium rolfsii.*
16. **Die-back:** *Diaporthe nomurai, Hendersonula toruloidea, Phoma sorghina.*
17. **Stem blight:** *Phloeospora ulmi, Phoma exigua.*
18. **Stem canker:** *Lasiodiplodia theobromae.*
19. **Stem rot:** *Claudopus nidulars, Ganoderma lucidum, Sclerotinia libertana, Tubercularia vulgaris.*
20. **Twig blight:** *Coniothyrium foedans, Fusarium lateritium, Fusarium moniliforme, Fusarium roseum, Fusarium solani, Myxosporium diedickii, Stigminia mori, Polyporus hispidus.*
21. **Root rot / wilt:** *Rhizoctonia bataticola* (= *Macrophomina phaseolina*)
22. **Cotton root rot:** *Phymatotrichum omnivorum*
23. ***Armillariella* root rot:** *Armillariella mellea*

24. Voilet root rot: *Helicobasidium mompa*

25. White root rot: *Rosellinia necatrix*

26. Bacterial leaf spot: *Xanthomonas campestris* pv. *mori* (Pommel) Dowson.

27. Bacterial leaf scorch: *Xylella fastidiosa*

28. Bacterial wilt: *Ralstonia solanacearum*

29. Bacterial rot I: *Bacterium moricolum*

30. Bacterial rot II: *Bacterium mori*

31. Soft rot: *Erwinia carotovora* subsp. *carotovora*

32. Yellow dwarf: *Candidatus* Phytoplasma

33. Mulberry latent virus

34. Mulberry ringspot virus

35. Tobacco necrotic virus

36. Mulberry yellow net vein virus

37. Potato spindle tuber viroid (Syn: *Mulberry mosaic dwarf viroid*)

Selected References

Biswas S, Singh NP. 2005. Parasitic Diseases of Mulberry, Kalyani Publishers, India.

Chowdhary SK. Raj SK. 1986. *Fusarium solani* (Mart.) Sacc., A new pathogen causing leaf spot disease in mulberry and its control. Sericologia, 26Q):125-130.

Dayakar yadav BR, Sukumar J, Prasad KV. 1993 Screening for potential resistance in the mulberry to leaf spot (*Cercospora moricola* cooke) disease. Sericologia, 33(1):81-90.

Dutta SK, Ghosh MK, Borah A, Bindroo BB. 2012. Development of disease forecasting model for leaf rust of mulberry (*Morus alba* L.) of Dimapur of North East India. Nature and Science, 10(8): 66-69.

Dutta SK, Maji MD, Gosh A, Roy Chowdhury S, Choudhuri B, Majumdar MK, Bindroo B. 2011. Survey on disease severity of mulberry (*Morus alba* L.) in different seasons of Aizawl districts (Mizoram). Journal of Crop and Weed, 7(2): 253-253.

Gosh, Lipika, Ahsan N, Parvez S, Swaraz AM, Khan MR, Ali MR, Alam MF. 2003. Survey and evaluation of leaf spot disease in six varities of mulberry (*Morus* sp). Journal of Biological Science, 3(12):1070-1075.

Govindadaiah, Gunasekhar V, Putta Sware Gowda and Thiagarajan V. 1994. Field evaluation of fungicides against *Phyllactina corylea* causing powdery mildew in mulberry. Indian Journal of Sericulture, 33: 160-162.

Govindaiah, Gargi, Sharma DD. 1987. *Furarium* diseases of mulberry. News Lettcr, Central Sericultural Research and Training Institute, Mysore, 2(3) Oct. pp.5

Govindaiah, Sengupta K. Sharma DD, Argi Gunasekhar V. 1989a. A new leaf spot disease of mulberry caused by *Myrothecium roridum* Tode ex Fr. Curr.Sci., 58(7):398.

Govindaiah, Sharma DD, Sengupta K, Gunasekhar V. Suryanarayana N, Madhav Rao, YR. 1989b. Screening of mulbcrry varieties against major fungal diseases. Indian J. Seric. 35; 207- 213.

Gunasekhar V, Philip Tomy, Govindaiah, Sharma DD, Nagaraj B, Datta RK. 1994. Seasonal occurrence of foliar fungal and bacterial disease of mulberry in South India. Indian phytopthalogy. 47(1): 72-76.

Illahi, Irfan, Mittal Vishal, Ramegowda GK, Dhar, Anil, Khan MA. 2011. Occurance of major foliar disease of mulberry under temperate climatic conditions of Kashmir, India. International Journal of Science and Nature, 2(1), 51-54.

Khan MA, Dhar A, Zeya SB, Trag AR. 2004. Pests and diseases of mulberry and their management. 40-41. Bishen Singh Mahendra Pal Singh, Dehradun.

Maji Manas Dev, Haradhan, Kumar Das, Bimal. 2009. Screening of mulberry germplasam lines against powdery mildew, *Myrothecium* leaf spot and *Pseudocercospora* leaf spot disease complex. Archives of Phytopathology and Plant Protection, 42(9), 805-811.

Pratheesh kumar PM, Maji MD, Gangwar SK, Das NK, Saratchandra B. 2010. Development of leaf rust (*Peridiopsora mori*) and dispersal of uredinospores in mulberry (*Morus* spp). International Journal of Pest Management, 46(3):195-200.

Sengupta K, Govindaiah, Pradipkumar. 1991. Diseases and Pests of Mulberry and their Control, Central Sericultural Research and Training Institute, Mysore, India.

Sengupta K, Pradip, Kumar, Baig, Govidaiah.1990. Hand book on pest and disease control of mulberry and silkworm. Bangkok: United Nations, ESCAP.

36 Diseases of *Peltophorum pterocarpum* (Copper pod)

S. No.	Disease	Pathogen
1.	Leaf blight	*Alternaria alternata*
2.	Root rot	*Ganoderma lucidum*

1. Leaf blight

Symptoms

- A brown water soaked lesion develops on the leaf and spreads fast to occupy major portion of the leaf leading to complete drying of the lamina.
- The disease can easily identified by presence of concentric rings in the blighted area showing the sporulation of the fungus.

Fungus: *Alternaria alternata*

Management

- Spray with Carbendazim + Mancozeb 0.2% twice at 15 days interval.

2. Root rot

Symptoms

- Pathogen enters through distal root end

- Pale leaves which later dries
- White mycelial strands seen in between barkand wood
- When most of roots are affected, trees are killed

Fungus: *Ganoderma lucidum*

Management

- Maintain the trees in healthy condition – application of manure.
- Do not wound the tree. If any wounds are there apply Bordeaux paste.
- Remove the sporocarp and destroy.
- Scrap the mycelia portion and apply Bordeaux paste.
- Remove severely affected trees.
- Isolate the affected tree from healthy trees.
- Drench the soil near the tree with 40 litres of 1% Bordeaux mixture at quarterly interval for thrice a year and repeat after 2-3 years.
- Root injection in the infected plant containing 2g of Aureofungin soil and 1 g of copper sulphate in 100 ml of water.

Selected References

Nagadesi PK, Arya A. 2013. Rotting of *Peltophorum ferrugineum* (Decne.) Benth. by pathogenic lignicolous fungi in Rajpipla, Gujarat, India. Journal on New Biological Reports. 2(1): 17–27.

Sahu RK, Agarwal VK. 2003. Effect of fungicidal seed treatment on seedborne fungi, germination and seedling vigour of rusty shield bearer (*Peltophorum ferrugineum* Decne-Benth.). J. Myco. and Pl. Path. 33(1), 84-88.

Shukla DN, Bhargava SN. 1978. Effect of moisture on seed fungal flora. Proceedings of the National Academy of Science India, 45, 199-200.

37 Diseases of *Phyllanthus emblica* (Aonla)

S. No.	Disease	Pathogen
1.	Rust	*Ravenelia emblicae, Phakopsora phyllanthi*
2.	Anthracnose	*Glomerella cingulata*
3.	Sooty mould	*Capnodium* sp.
4.	Blue mould	*Penicillium islandicum*
5.	*Phoma* dry fruit rot	*Phoma emblicae*
6.	*Phoma* fruit rot	*Phoma putaminum*
7.	*Phomopsis* fruit rot	*Phomopsis phyllanthi*
8.	*Nigrospora* fruit rot	*Nigrospora sphaerica*
9.	*Cladosporium* dry fruit rot	*Cladosporium tenuissimum, Cladosporium cladosporioides*
10.	*Pestalotia* fruit rot	*Pestalotia cruenta*
11.	*Alternaria* fruit rot	*Alternaria alternata*

1. Rust

Symptoms

» Reddish brown, circular, 2-5 mm dia, uredo pustules appear on both surfaces of the leaflets.

» The uredosori rupture the epidermis expose uredospores.

» On fruits, pustules are black at first and later they develop into rings.

» Later the pustules join together and cover a big area of the fruit.

» The black spores get exposed by rupturing a paper covering and the fruit gives a dirty appearance.

Fungi: *Ravenelia emblicae, Phakopsora phyllanthi*

Management

» Spray with Wettable sulphur 0.3% or Zineb 0.2%.

2. Anthracnose

Symptoms

» Initial symptom of the disease is in the form of minute, circular, brown to gray spots with yellowish margin on leaflets.

» The central area of spots remains grayish with dot like fruiting bodies.

» On the fruits, lesions are depressed and dark at the centre.

» Acervuli arranged in rings are found on the fruit spots.

» The fruits become shriveled and rot.

Fungus: *Glomerella cingulata*

Management

» Keep the orchard clean.

» Collect and destroy the affected fruits from the orchard.

» Spray with Carbendazim 0.1% before harvesting of fruits.

3. Sooty mould

Symptoms

» Sooty mould causes velvety covering of black fungal growth on the surface of leaves, twigs and flower.

- Infection is restricted only to the surface and do not penetrate into leaves

Fungus: *Capnodium* sp.

Management

- Spray with starch 2% + Lambda cyhalothrin 0.05% + Wettable sulphur 0.2%.

4. Blue mould

Symptoms

- It causes brown patches and water socked areas on the fruits.
- In severe case, three different types of colours develop viz., bright yellow, purple brown and bluish green.
- There is exudation of drops yellowish liquid on the fruit.
- The fruits emit a bad odor.
- The whole fruit finally gives a bluish brown pustulated and beaded appearance.

Fungi: *Penicillium islandicum, Penicillium funiculosum*

Management

- Careful handling of fruits.
- Sanitary conditions in storage and by gas treatment with NCl_3 and ozone.
- Treatment of fruits with borax and sodium chloride.

5. *Phoma* dry fruit rot

Symptoms

- The infected areas are light brown from 1.0 to 2.0 cm but subsequently they appear depressed and turn completely black.
- Occasionally pycnidia are observed on older parts of the infected fruits.

Fungus: *Phoma emblicae*

Management

» Careful handling of fruits.

» Sanitary conditions in storage and by gas treatment with NCl_3 and ozone.

6. Phoma fruit rot

Symptoms

» The rot starts as a small pinkish brown necrotic spot which extends towards both the ends of the fruits forming eye-shaped appearance.

» In severe cases lesions coalesce forming a bigger pustule.

» The mature lesions are dark brown in colour and severely infected fruits show wrinkling.

» The under lying tissues in the rotted fruits become soft.

Fungus: *Phoma putaminum*

Management

» Careful handling of fruits.

7. Phomopsis fruit rot

Symptoms

» Smoke brown to black round lesions develop within 2-3 days of inoculation.

» The diseased parts later show olive brown discolouration with water soaked area and cover the whole fruit within 8 days.

» The shape of the fruit is also deformed.

Fungus: *Phomopsis phyllanthi*

Management

» Keep the orchard clean.

» Collect and destroy the affected fruits from the orchard.

- Spray with Carbendazim 0.1% or Captafol 1.5% or Mancozeb 1% just after the harvest of fruits.

8. *Nigrospora* fruit rot

Symptoms

- Rot initiation is from a small dot of 2 mm diameter extending to a black ring spot upto 7 mm.
- Several rings adjacent to each other coalesce resulting in complete rolling of the infected fruits.

Fungus: *Nigrospora sphaerica*

Management

- Keep the orchard clean.
- Collect and destroy the affected fruits from the orchard.
- Spray with Carbendazim 0.1% or Captafol 1.5% or Mancozeb 1% just after the harvest of fruits.

9. *Cladosporium* dry fruit rot

Symptoms

Cladosporium tenuissimum

- The disease starts as colourless area, slightly soft and subsequently progresses in a circular manner.
- Light brown mycelial growth of the fungus is also evident on infected areas.
- The diameter of the lesions varies from 1.0 to 1.5 cm in 7 days.
- Pathogenicity of the fungus was proved. Injury is necessary for manifesting the infection.

Cladosporium cladosporioides

- The rot shows dark brown necrotic lesions.
- Slight growth of the organism appears in the necrotic cavity.

- The severity of infection is recorded only in matured and ripe fruits.
- Some freshly harvested fruits also show infection.
- One week old lesion measure, 0.7 to 1.2 mm in diameter

Fungi: *Cladosporium tenuissimum, Cladosporium cladosporioides*

Management

- Careful handling of fruits.
- Sanitary conditions in storage and by gas treatment with NCl_3 and ozone.

***10. Pestalotia* fruit rot**

Symptoms

- The spots on the fruits are mostly irregular and brown in colour.
- The disease usually starts as a brownish discolouration on the fruit surface, which develops slowly.
- Later, the spots become mummy brown and skin around them develops light brown colouration.
- At a relatively later stage the infected region becomes covered with white fluffy aerial growth of the fungus.
- The internal parts of the diseased fruit shows a dry dark brown area.

Fungus: *Pestalotia cruenta*

Management

- Keep the orchard clean.
- Collect and destroy the affected fruits from the orchard.

***11. Alternaria* fruit rot**

Symptoms

- The rot starts as a small brownish spherical necrotic spot which increases in circular fashion with the development of the disease.

- In advance stages the spots become dark brown to black and neighboring spots coalesce.
- The centre portion of the infected tissues becomes soft and pulpy.

Fungus: *Alternaria alternata*

Management

- Keep the orchard clean.
- Collect and destroy the affected fruits from the orchard.

Selected References

Choubey S. 2007. Phomopsis fruit rot of aonla and its management. M.Sc. (Agri.) Thesis, Anand Agriculture Univeresity, Anand.

Firminger TA. 1947. Firminger's Manual of Gardening for India. Thacke Spink Co. Ltd., Calcutta. Goyal RK, Patil RT, Kingsly ARP, Walia H, Kumar P. 2008. Status of post harvest technology of Aonla in India-A review. Am. J. Food Technol., 3: 13-23.

Jamaluddin A, Tandon M.P, Tandon R.N. 1975. A fruit rot of aonla caused by *Phoma*. Proc. Natl. Acad. Sci. India. 8(45):75–77.

Jamaluddin A. 1978. *Cladosporium* rot of fruits of *Phyllanthus emblica* caused by *Phomopsis phylanthi* punith and its chemical control. Nat. Accd. Sci. India. 5(6):183–185.

Jat RG, Goyal SK, Choudhary S. 2013. Management of *Aspergillus* and blue mould rot of anola fruits. Asian J. Bio. Sci. 8(1):91–93.

Kamthan KP, Misra R, Shukla AK. 1981. *Nigrospora* fruit rot of *Emblica officinalis*, a new disease record. Sci. Cult. 47:371–372.

Lal B, Arya A, Rai R.N. 1982. A new soft rot of aonla caused by *Phomopsis phyllanthi* Punith and its chemical control. Nat. Acad. Sci. Lett. 5(6):183–185.

Pandey RS, Bhargave SN, Shukla DN, Divedi DK. 1984. Two new fruit diseases of aonla caused by *Alternaria* sp. Int. J. Trop. Pl. Dis. 2:79–80.

Pandey RS, Shukla D.N, Khati DV, Bhargava SN. 1980. A new fruit rot of *Phyllanthus emblica*. Indian Phytopathol. 27:121–122.

Pathak PK, Preeti D, Kumar S. 2009. Effect of postharvest treatments on shelf life of aonla (*Emblica officinalis*) fruits damaged during harvesting. J. Food Sci. Technol. 46:283–285.

Rathod RS, Patel JG. 2004. Effect of pre and post inoculation treatment on post harvest rots of aonla fruits. J. Myco. Plant Pathol., 35: 525-525. Rawal RD, Saxena AK. 2005. Management of post harvest diseases of tropical fruits with reference to India. Proceedings of the 2nd Global Conference, Plant Health-Global Wealth, November 25-29, 2005, Udaipur, India, pp: 1-33.

Rawal RD. 1993. Fungal disease of tropical fruits spp. In : K. L. Chaddha and O. P. Pareek (Eds.). Advances in Horticulture, fruit crops, Vol. 3 Malhotra Publ. House, New Delhi, pp. 1255-1273.

Sengupta P, Sen S, Mukherjee K, Acharya K. 2019. Postharvest Diseases of Indian Gooseberry and Their Management: A Review, International Journal of Fruit Science, DOI: 10.1080/15538362.2019.1608889

Sharma N, Mashkoor Alam M. 1998. Post Harvest Diseases of Horticultural Perishables. International Books Distributing Co., Lucknow, pp: 226-253.

Thom C. 1930. Blue mold of aonla (*Phyllanthus emblica* L.). Mycologia 6:210–211.

Tyagi RNS, Pathak AK. 1988. Control of aonla (*Emblica officinalis* Gaertn.) rust (*Ravenelia emblicae* Syd.) through fungicides. J. Mycol. Pl. Pathol., 17: 218-219.

Verma LR, Sharma RC. 1999. Diseases of Horticultural Crops: Fruits, Indus Publishing Company, New Delhi.

38 Diseases of *Picea smlthiana* (Spruce fir)

S. No.	Disease	Pathogen
1.	*Leucostoma* canker	*Leucostoma kunzei*
2.	*Rhizosphaera* needle cast	*Rhizosphaera kalkhoffii*

1. *Cytospora* canker/ *Leucostoma* canker

Symptoms

» The disease normally starts on the lowest branches of the tree and, over a period of several years, progresses upward.

» At first, needles have a purplish hue, eventually turning brown and dropping, leaving dry, brittle twigs and branches.

» On severely infected trees, the fungus will enter the trunk through wounds (usually where the branch meets the trunk of the tree), killing the cambium layer and leaving dead bark.

» This dead tissue is called a "canker."

» A conspicuous white resin or "pitch" covers the cankered portion of the branch or trunk, sometimes flowing several feet down the trunk of the tree.

» This is an important means of diagnosing Cytospora canker; however, resin flow can also be associated with other tree injuries and is not exclusively symptomatic of Cytospora canker.

» Within the cankered area, black, pinhead-size fruiting structures (pycnidia) of the fungus can be seen with a microscope or hand lens and are a positive sign of the disease.

Fungus: *Leucostoma kunzei* (formerly *Cytospora kunzei*). Teleomorph is *Valsa kunzei*

Management

» Avoid injury in roots.

» Soil drenching with organic amendments.

2. Rhizosphaera needle cast

Symptoms

» Needles to appear mottled or speckled with dull yellow or reddish blotches.

» Later, (often the next year) infected needles on the interior of a branch turn purplish-brown (from the tips downward) and drop prematurely.

» Infection generally begins in spring on the needles of the lower branches soon after the needles have elongated.

» Symptoms spread upward and around the tree.

» Sometimes infection will start on branches in the middle of a tree, creating defoliation "holes" among healthy branches.

» Heavily infected trees can suffer severe needle loss and branches may die as they become defoliated.

» Trees are rarely killed by *Rhizosphaera* needle cast, but several years of attack will take its toll and only the current season needles may remain.

» In moist conditions, the fungus inside older needles produces black fruiting structures (pycnidia) that appear as distinct rows of black, pinhead-size dots.

» These fruiting structures emerge through needle pores (stomata), either before or after the needles have dropped.

» This can be seen with a hand lens or by the unaided eye.

» Healthy stomata appear white.

» The rows of black stomata are a diagnostic feature of Rhizosphaera needle cast.

» This disease can be frustrating because severe defoliation can occur quite rapidly and without indication that the disease is even present.

Fungus: *Rhizosphaera kalkhoffii*

Management

» Remove (when feasible) dead branches, fallen needles, and cones under the tree to prevent further infections.

» Prune surrounding plants to promote better air circulation and keep plants well watered, especially in periods of drought, to alleviate stress.

Other Diseases

3. **Red band**: *Mycosphaerella pini* (*Dothistroma septosporum*)

4. ***Elytroderma* needle cast**: *Elytroderma deformans*

5. **Spruce broom rust**: *Chrysomyxa arctostaphyli*

6. **Western gall rust**: *Endocronartium harknessii*

7. **Pinyon blister rust**: *Cronartium occidentale*

8. **Crown and collar rot**: *Phytophthora cactorum, P. megasperma, P. cambivora*

9. ***Armillaria* root rot**: *Armillaria ostoyae*

10. ***Schweinitzii* crown rot**: *Phaeolus schweinitzii*

11. **Black stain root**: *Leptographium wageneri*

Selected References

Hawksworth FG, Hinds TE. 1965. *Cytospora* canker of Engelmann spruce in Colorado. Plant Disease Reporter, 49: 72.

Proffer TJ, Hart JH. 1994. Wound inoculations of Colorado blue spruce with isolates of *Leucostoma* (*Cytospora*) *kunzei* and other *Cytospora* species. Journal of Arboriculture, 20: 215–221.

39 Diseases of *Pinus roxburghii* (Chir pine)

S. No.	Disease	Pathogen
1.	Damping off	*Fusarium* sp., *Rhizoctonia* sp., *Pythium* sp., *Phytophthora* sp.
2.	Pine *Swertia* felt rust	*Cronartium himalayense*
3.	*Coleosporium* needle rust	*Coleosporium campanulae*
4.	*Pseudocercospora* needle blight	*Pseudocercospora pini-densiflorae*

1. Damping off

Symptoms

» The seedlings are prone to attack at different stages of growth and accordingly the disease is termed as pre-emergence and post emergence damping off and root rot.

» Seed decay before it emerges from the soil (pre-emergent damping-off).

» Seedlings may be killed after they emerge from the soil (post-emergence damping-off).

» Water soaked lesions and browning at seedlings at collar region, root decay

» Death of seedlings in patches.

Fungus: *Fusarium* sp. and *Rhizoctonia* sp.

Oomycete: *Pythium* sp., *Phytophthora* sp.

Management

- Avoid water logging, prepare raised beds.
- Seed treatment with captan 2 g/ kg seeds *Pseudomonas fluorescens* at 10 g/ kg seed.
- Avoid wounding plants as these will allow entry of the fungus.
- Adequate spacing to avoid crowding and the formation of high humidity.
- Do not overwater as disease is more severe in very wet soils.
- Remove affected seedlings.
- Apply *Pseudomonas fluorescens* at 2.5 kg/ 50 kg FYM to nursery bed.
- Soil drench with Copper oxy chloride 0.05% or Carbendazim at 0.1%.
- Soil drench with Blue copper 0.05% or Brassicol or Captan 0.1%.

2. Pine *Swertia* felt rust

Symptoms

- Chir pine is susceptible to infection up to 10-15 years.
- Initially infected portion exude resin like substances.
- Later, bark cracking and canker develop, where small pycnia and aecial cups burst through the ruptured bark.
- Infected branches become pale in the needles and killed.

Fungus: *Cronartium himalayense* the peridermium stage is *Peridermium himalayense* (Syn: *Peridermium pini*)

Two hosts: Macrocyclic heterocious rust

- Pycnia and aecia on Chir pine
- Uredia and telia on Swertia

Chir pine: Pycnia appear on bark of chir pine as minute blisters which release pycniospores as orange yellow fluid. Aecia appear on stems as blisters from march to June where orange aeciospores are liberated to long distances on Swertia

Swertia: Telia develop on swertia which release light brown cylindrical teliospores with felty appearance. Later teliospores produce promycelium that gives rise to basidiospores that in turn affects chir pine.

Management

» Eradication of Swertia weed from pine forests (up to 300 m) for three years.

3. *Coleosporium* needle rust

Symptoms

» The infection occurs all ages and stands, especially abundant on the dense and bushy foliage of young trees.

» The affected needle turns pale yellow, around the sori and shed.

Fungus: *Coleosporium campanulae*

Macrocyclic heteroecius rust

Two hosts: Chir pine and Campanula

» Pycnia and aecia on Chir pine

» Uredia and telia on Campanula

Chirpine: Aecia develops release aeciospores and infects Campanula

Campanula: Uredia develop on leaves, stems and produce urediniospores. Telia develop and produce teliospores that germinate to produce promycelium that produce sporidia which infects Campanula

Management

» Eradication of Swertia weed from pine forests (up to 300 m) for three years.

4. *Pseudocercospora* needle blight

Symptoms

» The pathogen infects needles producing pale green lesions which later turn brown to grayish brown.

» The disease manifests in the lower needles and later progresses upwards.

Fungus: *Pseudocercospora pini-densiflorae*

Management

» Spray Mancozeb 0.5 g / lit at triweekly intervals.

Other Diseases

5. **Root rots:** *Armellarea mellea, Fomes annosus, Ganoderma lucidum, Polyporus schweinitzii, Helicobasidium compactum, Fusarium oxysporum*

6. **Heart rots**: *Fomes pini, Fomes pinicola, Ganoderma applanatum, Poria monticola*

7. **Needle blight**: *Dothistroma pini, Pestalotia funerea, Cercospora pini-densiflorae, Trimmatostroma* sp.

8. **Twig blight**: *Diplodia pinea*

9. **Diplodia tip blight, Diplodia dieback, Diplodia shoot blight**: *Diplodia* sp.

10. **Pitch canker, Stem rot, Root rot**: *Rhizoctonia* spp.

11. ***Pestalotiopsis* foliage blight, canker, dieback, needle disease, tip blights, grey blights and chlorosis**: *Pestalotiopsis* spp.

12. **Needle cast**: *Lophodermium pinastri*

13. ***Cercoseptoria* needle blight**: *Cercoseptoria pini*

14. **Pitch canker**: *Fusarium circinatum*

15. **Sooty mould**: *Capnodium pini*

16. **Flowering parasite**: *Dendrophthoe falcata*

Selected References

Brown A, Green S, Hendry S. 2005. Needle diseases of pine. Forestry Commission Information Note 67. Forestry Commission, Edinburgh.

Carlucci A, Colatruglio L, Frisullo S. 2007. First report of pitch canker caused by *Fusarium circinatum* of *Pinus halepensis* and *P. pinea* in Apulia. Pl. Dis. 91:1683

Gupta B, Dass B. 2007. Composition of herbage in *Pinus roxburghii* Sargent stands: basal area and Importance value index. Caspian J. Environmental Sciences 5: 93–98.

Huang JW, Kuhlman EG. 1990. Fungi associated with damping off of slash pine seedlings in Georgia. Plant Dis. 74:27-30.

Siddiqui MF. Ahmed M, Wahab M, Khan N. 2009. Phytosociology and structure of *Pinus roxburghii* Sargent (Chir Pine) in lesser Himalayan and Hindukush range of Pakistan. Pak. J. Bot. 41: 2357–2369.

40 Diseases of *Pinus strobus* (White Pine)

S. No.	Disease	Pathogen
1.	Die back	*Lecanosticta acicula*, *Lophophacidium dooksii*, *Bifusella linearis*, *Septorioides strobe*
2.	White pine root decline	*Leptographium procerum*
3.	Root rot	*Phytophthora* sp.

1. Die-back

Symptoms

- » Needles of mature trees become straw-colored to brown before they are prematurely shed from the canopy.
- » In some cases, only a few main branches are symptomatic whereas on other trees, the entire canopy exhibits the symptoms.

Fungi: *Lecanosticta acicula* (formerly *Mycosphaerella dearnessii*), *Lophophacidium dooksii* (formerly *Canavirgella banfieldii*), *Bifusella linearis*, *Septorioides strobe.*

Management

- » Removal and destruction of infected bushes.

2. White pine root decline

Symptoms

- Early symptoms include delayed bud break and reduced candle elongation in spring; however, these subtle symptoms are often overlooked.
- As root rot progresses, mature foliage begins to fade, droop, and turn brown in stark contrast to healthy trees nearby.
- Chocolate-colored cankers can form at the base of affected trees and cause dieback.
- Once needle browning is clearly visible, needle browning advances and tree death occurs quickly.
- Resin is visible on bark at the tree base where a flattened canker develops.
- Wood beneath bark becomes discoloured.
- The causal fungus penetrates inner bark and sapwood of roots and lower trunk of white pine.
- Fungal spores may build up in soil around infected roots and spread from tree to tree via soil water or runoff. Thus, trees planted in wet sites may seem to be more susceptible to white pine root decline.
- Disease can also occur in moderately dry or well-drained soil. Bark-infesting insects and other fungi such as those causing "sapstain" or "bluestain" (internal discoloration) are also commonly present in declining trees.
- Spread may occur as contaminated insects emerge from diseased trees and feed on healthy trees nearby.

Fungus: *Leptographium procerum* (*Verticicladiella procera*)

Management

- Remove and destroy infected trees including stumps.
- Do not replant white pine among the stumps of recently killed trees.

3. Root rot

Symptoms

- Trees infected with *Phytophthora* root rot have soft, brown decaying roots, while above-ground symptoms include stunted, yellow needles.
- Decay may also be evident in lower trunks as reddish-brown discoloration beneath the bark. Some trees may become stunted, wilt, or gradually exhibit dieback.
- In contrast, others (typically younger trees) may turn uniformly brown and die rapidly.

Oomycete: *Phytophthora* sp.

Management

- Plant white pine (and other susceptible plants) in well-drained sites, or improve drainage by diverting water flow, tiling, installing raised plantings, etc.
- Soil drenches with Copper oxychloride 0.25%.

4. Stem blister rust: *Cronartium ribicola*

Selected References

Banfield WM. 1960. *Lophodermium* needle cast of the eastern white pine. Phytopathology, 50, 268.

Banfield WM. 1962. Experimental evidence of pathogenicity for the *Lophodermium* associated with dead needles of eastern white pine. Phytopathology, 52, 2.

Bebber DP, Ramotowski MAT, Gurr SJ. 2013. Crop pests and pathogens move polewards in a warming world.

Boyce JS Jr. 1959. Brown spot needle blight on eastern white pine. Plant Disease Reporter, 43, 420.

Broders KD, Wyka S, Munck IA. 2014. Characterization and distribution of fungi associated with needle defoliation of eastern white pine (Pinus strobus). International Forestry Review, 16, 322.

Broders KD, Munck I, Wyka S, Iriarte G, Beaudoin E. 2015. Characterization of fungal pathogens associated with white pine needle damage (WPND) in northeastern North America. Forests, 6, 4088–4104.

Horst ZY, Westcott C. 2008. Westcott's Plant Disease Handbook, 7th edn. Springer, Berlin, New York.

Huang ZY, Smalley EB, Guries RP. 1995. Differentiation of *Mycosphaerella dearnessii* by cultural characters and RAPD analysis. Phytopathology, 85, 522–527.

Ingold CT. 1971. Fungus Spores: Their Liberation and Dispersal. Clarendon Press, Oxford.
Ingold CT. 1978. Role of mucilage in dispersal of certain fungi. Transactions of the British Mycological Society, 70, 137–140.

Merrill W, Wenner NG, Dreisbach TA. 1996. *Canavirgella banfieldii* gen. and sp.nov.: a needlecast fungus on pine. Canadian Journal of Botany, 74, 1476–1481.

Minter DW, Millar CS 1984. *Bifusella linearis*. IMI descriptions of fungi and bacteria. CABI Bioscience, Bakeham Lane, Egham, Surrey, TW20 9TY, UK

41 Diseases of *Pinus wallichiana* (Blue Pine)

S. No.	Disease	Pathogen
1.	Heart rots	*Fomes pini, Fomes pinicola, Fomes rosues, Fomes fomentarius, Fomes conchatus, Polyporus gilvus, Polyporus dryadeus, Polyporus cuticularis*
2.	Stem blister rust	*Cronartium ribicola*
3.	Needle rusts	*Coleosporium himalayense, Coleosporium barclayense, Melampsora oblonga, Pucciniastrum agrimoniae, Pucciniastrum aceris*
4.	*Lophodermium* needle blight	*Lophodermium macci*
5.	*Myrothecium* needle blight	*Myrothecium verrucaria*
6.	Dwarf mistletoe	*Arceuthobium minutissimum, Arceuthobium oxycedri, Arceuthobium sichuanense*

1. Heart rots

Symptoms

- It is a white rot fungus causing heart rot.
- Incidence more in higher rainfall areas.
- Fungus enters through branch stubs and open knots.
- Coppicing of branches invites infection, wounds, infection through root grafts causes

decay in heart wood which becomes stained light pink/ brown / red

» White pockets develop in heart wood.

» Sporophores develop on stumps upper surface greyish black, cracks and zones.

Fungi: *Fomes pini, Fomes pinicola, Fomes rosues, Fomes fomentarius, Fomes conchatus*

Other Fungi: *Polyporus gilvus, Polyporus dryadeus, Polyporus cuticularis*

Management

» Control coppicing.

2. Stem blister rust

» The fungus attacks young plants/ branches which become swollen, blistered and, girdled.

» Resin exudation occurs.

» Macrocyclic heteroecious rust.

Fungus: *Cronartium ribicola*

Two hosts

Blue pine: Pycnia and aecia

» Pycnia are minute on blue pine that release pycniospores

» Aecia – emerge on bark in spring and release aeciospores. The aecial stage is named as *Peridermium indicum*

Ribes rubrum: Uredia and telia

» Uredia release spiny orange yellow uredeospores

» Telia develop and release - develop in to promycelium which release sporidia that infects blue pine

Management

» Eradication of *Ribes rubrum*

3. Needle rusts

***Coleosporium* needle rust**	***Melampsora* needle rust**	***Pucciniastrum* needle rust**
Coleosporium himalayense *Coleosporium barclayense*	*Melampsora oblonga*	*Pucciniastrum agrimoniae* *Pucciniastrum aceris*
Two hosts: Blue pine and Senecio rifinervis	-	Two hosts: Blue pine and *Agrimonia acer*
Macrocyclic rust	Microcyclic lepto-form rust	Microcyclic rust
Pycnia are dark brown release pycniospores. Aecia are needle like and release orange red round aeciospores	Not known	Not known
Uredia develop on *Senecio rufinervis* and release oval spiny uredeospores. Telia release teliospores, hyaline single celled produce promycelium develops sporidia infects blue pine	Only teliospores known on blue pine. Telia are pinkish red, release hyaline single celled teliospores produce promycelium develops sporidia infects blue pine	Uredospores are obovoid and echinulated. Telia is small, reddish brown pads that produce.

4. Lophodermium needle blight

Symptoms

» The disease on needles initially appeared as yellowish spots which later turned brown with yellowish halos.

» The spots on needle later coalesced to give blighted appearance and ultimately resulted in premature defoliation.

» The stereoscopic examination of blighted needles showed the presence of small pycnidia (spermatia).

» The disease reported on *Pinus wallichiana* and *Pinus halepensis.*

Fungus: *Lophodermium macci*

Management

» Removal and destruction of infected needles and bushes.

- Spray Bordeaux mixture 1%.

5. *Myrothecium* needle blight

Symptoms

- First observed as slight whitish chlorotic lesions of 0.1 to 0.5 mm size on needles.
- Then these lesions enlarged to 1.0– 1.8 mm in size and turned greenish-yellow in colour.
- Later on, the centre of these lesions turned straw to brown with dark brown margins of 0.8 mm width
- An erumpent creamy fungal layer appeared on these brown spots with oval to irregular sporodochia formed.
- Finally, these lesions elongated parallel to the length of needles, with browning in both directions from the infection site and premature needle defoliation in severe cases.

Fungus: *Myrothecium verrucaria*

Management

- Removal and destruction of infected needles and bushes.
- Spray Bordeaux mixture 1%.

6. Flowering parasite (Dwarf mistletoe)

- Occurs in low rainfall areas
- Infection noticed as spindle shaped or fusiform swellings on branches of main bole results in top drying, death of trees
- The parasite produces male and female flowers, fast spread
- Infection spreads into berries; seeds ejected to long distance and adhere to stems of blue pine through mucilage.

Parasite: *Arceuthobium minutissimum*, *Arceuthobium oxycedri*, *Arceuthobium sichuanense*

Management

» Pruning of infected branches.

» Apply bordeaux paste.

» Removal of berries.

Other Diseases

7. **Seedling blight**: *Sclerotinia fuckeliana*

8. **Root rot**: *Phytophthora cinnamomi, Fusarium oxysporum* f.sp. *pini, Rhizoctonia solani, Macrophomina phaseolina*

9. ***Annosum* root disease**: *Heterobasidium annosum*

10. **Pitch canker**: *Fusarium subglutinans*

11. **Rust**: *Cronartium quercum* f. sp. *fusiforme*

12. ***Diplodia* tip blight**: *Sphaeropsis sapinea*

13. **Red band needle blight**: *Scirrhia pini, Dothistroma septosporum, Dothistroma pini*

14. ***Myrothecium* needle blight:** *Myrothecium roridum*

15. ***Lophodermium* needle cast**: *Lophodermium seditiosum, Lophodermium pinastri*

16. ***Ploioderma* needle cast**: *Ploioderma* spp.

17. ***Cercospora* needle blight**: *Cercospora pinidensiflora, Mycosphaerella gibsonii*

18. ***Meloderma* needle blight**: *Meloderma desmazierii*

19. ***Phoma* blight**: *Phoma eupyrene*

20. **Leaf blight**: *Botryosphaeria obtusa*

21. **Needle cast**: *Hypoderma desmazierii, Dothistroma pini, Leptostroma ahmadi, Leptostroma pinastri, Leptostroma pini-excelsa*

22. **Sooty mould**: *Capnodium pini*

Selected References

» Bagchee K, Singh U, 1960. List of common names of fungi attacking Indian forest trees, timber and the herbaceous and shrubby undergrowths and list of cultures of forest fungi. Indian For. Rec., 1: 199-348.

» Barnard EL, Ash EC. 1998. Needle cast of pines in Florida. Plant Pathology Circular No. 368, pp: 1-4.

» Bilgrami KC, Jammaludin, Rizvi MA. 1991. The Fungi of India. Today's and Tomorrow's Publishers, New Delhi, India, Pages: 278.

» Farooq A. Ahanger, Hassan Dar, Muzafar A, Beig, Sofi TA. 2011. First report of needle blight on blue pine (*Pinus wallichiana*) and Aleppo Pine (*P. halepensis*), caused by *Lophodermium macci*, from Asia. Plant Pathology Journal, 10: 181-186.

42 Diseases of *Populus* sp. (Poplar)

1. Bipolaris blight

Symptoms

» Pink to brown, pin head size spots develop on leaves.

» Gradually, these enlarge into dark brown, irregular spots, often surrounded by a chlorotic margin.

» In rainy or humid weather, the irregular spots coalesce to cover the leaf blade that gave blighted appearance to the entire foliage.

» Sometimes, the chlorotic spots were marginalized through midrib and veins giving a defined pattern of disease spread.

» Severely infected leaves crumble in dry weather and ultimately fall off prematurely.

» Necrotic lesions, sometimes developing into sunken, black spot also appear on green pedicles and shoots.

» Premature defoliation of the seedlings took place in short span of 3-4days.

Fungus: *Bipolaris maydis, Bipolaris setariae*, *Bipolaris specifera*

Management

» Spray with Carbendazim 0.2% or Mancozeb 0.2%.

Other Diseases

2. ***Botryodiplodia* sett rot:** *Botryodiplodia palmarum* (*Lasiodiplodia palmarum*), *B. palmarum*
3. **White root rot**: *Dematophora necatrix, Rosselinia nectrarix, Macrophoma* sp.
4. ***Ganoderma* root rot**: *Ganoderma lucidum*
5. ***Rhizoctonia* root rot:** *Rhizoctonia solani*
6. ***Poria* root rot:** *Poria vincta*
7. **Stem rot**: *Phellinus phephyphloccus, Phellinus noxius*
8. **Heart rot**: *Trametes* sp
9. **Leaf rust**: *Melampsora ciliate, Melampsora larici-populina* and *M. populina* on *P. alba*
10. ***Alternaria* leaf spot and twig/ tip blight**: *Alternaria alternata*
11. **Die back**: *Cytospora chrysosperma*
12. ***Cladosporium* leaf spot**: *Cladosporium humile*
13. **Powdery mildew**: *Uncinula solicis, Oidium sp.*
14. ***Cercospora* leaf spot**: *Cercospora populina*
15. ***Alternaria* tip blight**: *Alternaria* stage of *Pleospora infectoria*
16. ***Septoria* leaf spot**: *Septoria populi*
17. ***Sphaceloma* leaf spot**: *Sphaceloma populi*
18. ***Myrothecium* leaf spot**: *Myrothecium roridum*
19. ***Phyllosticta* leaf spot**: *Phyllosticta adjuncta*
20. ***Sclerotium* leaf spot**: *Sclerotium rolfsii*
21. ***Dreshclera* leaf spot**: *Dreshclera maydis*
22. ***Phomopsis* leaf spot**: *Phomopsis sp.*
23. ***Curvularia* leaf spot**: *Curvularia lunata*

24. ***Asteroma* leaf blight**: *Asteroma frondicola*

25. **Leaf blight**: *Bipolaris maydis, B. setariae, B. specifera*

26. ***Taphrina* leaf blister**: *Taphrina aurea*

27. ***Pollacia* blight**: *Pollacia elegans, Pollacia nigra*

28. **Pink disease**: *Phanerochaete salmonicolor*

29. **Bark burst**: *Botryodiplodia palmarum*

30. ***Fusarium* rot, wilt, leaf spot**: *Fusarium incarnatum equiseti* complex, *Fusarium semitactum, Fusarium solani*

31. ***Nigrospora* rot**: *Nigrospora* state of *Khuskiya oryzae*

32. **Defoliation**: *Drepanopeziza populorum*

33. **Canker and die-back**: *Cryptodiaporthe populea*

34. **Leaf spot and Canker**: *Davidiella populorum*

35. **Mistletoe**: *Viscum album*

Selected References

Bagchee K. 1960. Pathology of Kashmir Forest - A survey and review of the problems and methods of protection. Dehradun, FRI and C. 85p.

Bakshi BK, Singh S. 1967. Rust on Indian forest trees. Indian Forster Records (NS), 2(8): 139-204.

Bakshi BK. 1976. Forest pathology: Principles and practice in forestry. Delhi, Controller of publications. 400p.

Bakshi BK. 1971. Indian Polyporaceae on trees and timber. New Delhi, Indian Council of Agriculture Research. 46p.

Dickinson CH, Lucas JA. 1982. Plant pathology and Plant Pathogens. 2nd ed. Oxford, Blackwell Scientific Publications. pp. 51-53.

Khan SN, Misra, BM. 1989. Pollaccia blight of poplars in India. Europian Journal of Forest Pathology, 19: 379-81.

Khan SN. 1988. Diseases of poplars in nurseries of Himachal Pradesh. In: Meeting of FRI Scientists with H.P. Forest Officers, Shimla, September 1988. Proceedings.

Khan SN. 1999. Principal diseases of poplars in India and their management. In: National Seminar on Poplar, Dehradun, 25-27 November 1999. Proceedings. Dehradun, Forest Research Institute. pp. 1-9.

Mishra S. 2011. Disease status and their management of poplar nurseries in Northern India. Ph.D. thesis. Forest Research Institute Deemed University, Dehradun. 148p.

Misra BM, Khan SN. 1991. Sclerotium leaf spot and stem rot of poplars. Indian Forester, 117 (8): 618-20.

Naraina U. 1988. Some additions to Indian Cerosporaea. Farm Science, 1(3): 36-39.

Pandey PC, Khan SN 1992. Disease status of poplars in India. In: 19th Session of the International Poplar Commission, Zaragoza, 22-25 September 1992. Proceedings. IPC. pp. 1- 7.

Panwar P. 2010. Studies on leaf spot disease (*Curvularia* sp.) of poplar. M.Sc. thesis. H.N.B. Garhwal University, Srinagar. 59p.

Rawat BS. 1981. Pest and disease of poplar plantations in U.P. Tarai. In: Symposium on Silviculture, Management and Utilisation of Poplars, Srinagar, 15-18 October 1979. Proceedings, edited by R.V. Singh. Delhi, Controller of Publications. pp. 126-129.

Rehill PS, Khan SN, Tiwari RK, Rawat, DS, Misra BM. 1988. Control of poplar rust *Melampsora ciliate* in nurseries. Indian Journal of Forestry, 2(1): 77-79.

Rehill PS, Puri SN. 1980. Some important diseases of poplars. Indian Journal of Forestry, 2: 176-83.

Sharma RC, Sharma S, Khurana DK. 1999. Assessment of poplar germplasm for resistance against *Alterneria alternata* causing leaf spot of poplar in Himachal Pradesh. Indian Forester, 125: 784-787.

Sharma RC, Sharma S. 2000. Status and distribution of foliar disease of poplar in Himachal Pradesh. Indian Phytopathology, 53 (1): 57-60.

Sharma RC. 2009. Poplar diseases in India and their management. ENVIS Forestry Bulletin, 9(1): 87-99.

Singh P, Singh S. 1975. Inscet pest and diseases of poplars in developing countries. In: Second World Technical Consultations on Forest Diseases and Insects, New Delhi, 7-12 April 1975. Proceedings.

Singh P, Singh S. 1986. Insect-pests and diseases of poplars. Dehradun, Forest Research Institute. 74p.

Singh S, Khan SN, Mishra BM. 1983. Some new and noteworthy disease of poplar in India. Indian Forester, 109(9): 636-644.

Singh S, Pandey PC, Singh A, Bahadur R, Mukharji SN. 1991. Studies on foliage diseases of poplar. Indian Forester, 117: (2): 135-142.

Singh S, Pandey PC, Singh A, Bahadur R, Mukherjee SN. 1991. Studies on foliage disease of poplars. Indian Forester, 117: 135-142.

Swati. 2009. Studies on leaf spot disease (*Phoma/Phyllosticta* sp.) of poplar. M.Sc. thesis. Ch. C.S. University, Meerut.70p.

43 Diseases of *Prosopis juliflora* (Prosopis)

S. No.	Disease	Pathogen
1.	Root rot	*Fusarium* sp.
2.	Collar rot	*Macrophomina phaseolina*
3.	Stem canker	*Botryodiplodia theobromae*
4.	Leaf blight	*Septoria prosopides*
5.	Twig blight and die-back	*Diplodia prosopides*

1. Root rot

Symptoms

» The symptoms were wilting of the leaves, diseased seedlings could be uprooted easily due to the rotting of the roots, and the phloem region could be peeled off easily with xylem vessels appearing black in colour.

Fungus: *Fusarium* sp.

Management

» The disease can be controlled by adopting silvicultural practices for avoiding excessive humidity in nursery beds.

» Soil drenching with Carbendazim 0.1% at monthly intervals.

2. Collar rot

Symptoms

- This disease is being reported for the first time on *P. juliflora* and was observed in 6-9 month old seedlings.
- Appearing in dry conditions, the fungus attacks the collar region, which eventually becomes black due the accumulation of black pycnidia and the infected seedling eventually dies.

Fungus: *Macrophomina phaseolina*

Management

- No control measures were identified.

3. Stem canker

Symptoms

- The canker expands and the quality of the sapwood deteriorates.
- Callus formation was observed on affected parts and secondary infection of stem borers and termites was also observed on dead parts.
- The disease can be minimised by reducing excessive lopping and other physical injuries.

Fungus: *Botryodiplodia theobromae*

Management

- No control measures were identified.

4. Leaf blight

Symptoms

- This appears as light yellow to green coloured lesions on the leaflets, leading to complete defoliation in severe stages of infection.

Fungus: *Septoria prosopides*

Management

» A foliar spray of copper based fungicides

5. Twig blight and dieback

Symptoms

» Young shoots usually injured by shoot borers were found to be susceptible to this pathogenic fungus, and the attack continues from the top of the tree in a downwards direction.

Fungus: *Diplodia prosopides*

Management

» A foliar spray of copper based fungicides.

Minor Diseases

6. **Anthracnose**: *Colletotrichum capsici*, *Colletotrichum cymopsicola*

7. **Leaf spot**: *Pleospora scirrhioides, Septoria prosopodis*

8. **Leaf blight**: *Ravenelia spicigera*

9. **Die back**: *Dalbargia dalbergiae, Dalbargia prosopidina, Nectria flavistroma, Nectria purtonii*

Selected References

Bakshi BK, Reddy RMA, Puri YN, Singh S. 1972. Forest Disease Survey. Final Technical Report Pt-480. Forest Research Institute. Dehra Dun, India. 117p.

Bakshi BK. 1977. Disease Insect Survey Report. Forest Research Institute, Dehra Dun, India. 46p.

Bilgrami KS, Jamaluddin and Rizwi MA. 1979. Fungi of India. Volume I. Today and Tomorrow Printers and Publishers. New Delhi, India.

Bilgrami KS, Jamaluddin and Rizwi MA. 1981. Fungi of India. Volume II. Today and Tomorrow Printers and Publishers. New Delhi, India.

Srivastava KK, Kalyani KB and Rajarishi R. 1989. Some noteworthy diseases of *Acacia nilotica* from south India and their management. Proceedings of a Seminar on Forest Production, Forest Research Institute. Dehra Dun, India.

44 Diseases of *Pterocarpus marsupium* (Indian Kino Tree)

1. **Defoliation:** *Marvalia pterocarpi*
2. **Root rot:** *Ganoderma lucidum*
3. **Heart rot:** *Polyporus gilures* and *Fomes fastuosus*
4. **Leaf spot:** *Cercospora cenescens*
5. **Leaf blight:** *Sclerotium rolfsii*, *Phomopsis* sp.
6. **Heart wood decay:** *Hymenochaete rubiginosa*

Selected Reference

Bakshi BK. 1976. Forest pathology principles and practices in forestry, Govt. of India Press, FRI and colleges, Dehradun, 400p.

45 Diseases of *Pterocarpus santalinus* (Red sandalwood)

1. **Leaf spot:** *Cylindrocladium scoparium*
2. **Leaf blight:** *Sclerotium rolfsii*, *Phomopsis* sp.

Selected Reference

Bakshi BK. 1976. Forest pathology principles and practices in forestry, Govt. of India Press, FRI and colleges, Dehradun, 400p.

46 Diseases of *Santalum album* (Indian sandalwood)

S. No.	Disease	Pathogen
1.	Damping off	*Rhizoctonia solani, Pythium* sp., *Phytophthora* sp.
2.	Wilt/ Canker	*Fusarium oxysporum*
3.	Root rot	*Phellinus noxius*
4.	Anthracnose	*Colletotrichum gloeosporioides*
5.	Powdery mildew	*Pseudoidium santalacearum*
6.	Spike	*Candidatus* Phytoplasma
7.	Giant-mistle toe	*Dendrophthoe falcata*

1. Damping-off and seedling wilt

Symptoms

» Death of seedlings was recorded in all the stages.

» Pre-emergence seed blight.

» Post-emergence mortality of seedlings.

» Root rot and wilting of seedlings.

Fungus: *Rhizoctonia solani*

Oomycete: *Pythium* sp., *Phythophthora* sp.

Management

- Prepare raised bed nursery.
- Avoid water stagnation.
- Soil drenching with FYM amended with bioconsortia of *Pseudomonas fluorescens* and *Trichoderma harzianum.*
- Soil drenching with copper fungicide and nematicide.

2. Wilt/ canker

Wilt symptoms

- Initially infected plants show yellowing.
- In seedlings infection, dwarfing, stunting and necrosis are common.
- Loss of turgidity leads to withering and drooping of leaves.
- Simultaneously, vascular plugging leads to death of plants.
- Infected tissues shows reddish color discoloration.

Canker symptoms

- The collar region of the dead plants showed bulged canker-structure.
- Almost all the plants in that patch were attacked with this problem.
- Probably water logging conditions under the canopy of a tree would have favored the infestation resulting in death of trees.

Fungus: *Fusarium oxysporum*

Management

- Soil drenching with FYM amended with bioconsortia of *Pseudomonas fluorescens* and *Trichoderma harzianum.*

3. Root rot

Symptoms

- The leaves of affected trees will rapidly turn from green to brown before dropping within a few weeks of showing the first symptoms.
- In the diseased area, a brown skin, which is the fruiting body of the fungus, can be seen at the base of the tree.
- The disease is spread through the root systems of affected trees, and infection can therefore rapidly spread to other trees in a plantation.

Fungus: *Phellinus noxius*

Management

- Avoiding planting in areas where the disease is already established.
- Selecting a site that is free-draining and on a slight slope; waterlogged soil promotes the growth of the fungus.
- Removing all living stumps when clearing a new site, because living stumps and roots are the most likely sites for *Phellinus* spores to establish a new colony; stumps should be burned to ensure that they are completely dead
- Planting herbaceous plants adjacent to each sandalwood tree (e.g. *Euphorbia tithymaloides* [vinil], *Alpinia purpurata* [red ginger] and *Cordyline fruticosa* [cordyline or nangaria] are said to help prevent.
- Avoiding unnecessarily cutting into the sandalwood tree with a bush knife pruning during dry conditions to limit the chance of infection of cut stems and to help the rapid healing of the wound.
- Removing and burning any fallen branches from the plantation.

4. Anthracnose

Symptoms

- It mainly occurs in the plant leaves, it's often damage leaf margin and tip, serious, most part of leaf becoming black and the dead.
- Produce round or nearly round lesion in the stem, there is a black dot in the pattern with pale brown.
- Lesion occurred on the tender shoots with oval ulcer plaque, margin slightly uplift.

- Sandalwood blight disease caused by anthracnose and harms the tender leaves, flowers, fruit and shoot of sandalwood.
- It mainly harms the tender leaves and fruit of sandalwood.
- Leaves will appear purple or will have purple-brown spots in the early stage of the disease.

Fungus: *Colletotrichum gloeosporioides*

Management

- Spray with Chlorothalanil 0.1% or Benomyl 0.25%.

5. Powdery mildew

Symptoms

- It mainly occurs in plants at seedling stage.
- It began to produce yellow dots on the leaves, and then, expanded into a round or oval spots with white powdery mildew layer on the surface.
- In general, the lower part of the blade is more than the upper leaves, and the back of the blade is more than the front.
- Mildew disperses individually in the early stage, and unites into a large blotch later, which can even cover the whole leaf.

Fungus: *Pseudoidium santalacearum* (Syn: *Oidium santalacearum*)

Management

- Spray with Triademefon 0.1% or Wettable sulphur 0.1%.

6. Spike

Symptoms

- The infected trees exhibit shortened internodal length leading to crowding of the leaves at the top.
- A drastic reduction in the size of the leaves was also observed.

- The leaves show tendency to stand stiff from the branches giving a spiky appearance.
- Later the leaves become yellowish and finally turn to red in colour.
- Spiked plants do not bear flowers or fruits and trees generally die within 1 to 2 years after the appearance of symptoms at a young age.
- Infected inflorescence produce only fruits with aborted seeds.
- In severe cases, the floral parts are modified into green leafy structures called 'Phylloids'.
- A large number of pleomorphic bodies are found to present in the infected tissues.
- The pathogen is transmitted through the leaf hoppers - *Jassus indicus* and *Moonia albimaculata*

Bacterium: *Candidatus* Phytoplasma

Management

- Managing insects by foliar spraying of Imidacloprid at 0.1%

7. Flowering parasite

Symptoms

- It produces a very hard flattened prostrating stem which run over the stem portions of the host and produce haustorial connections to drive the food source.
- The point of parasitism becomes thickened to form a gall like structure from which multiple shootings ranging from 15-20 arise as a clusters.
- In the case of the severe parasitism, the affected branch start withering drying from tip and dry off.
- The parasite is a shrubby plant with green leaves and produce erect, clustered branches from the galled portion.
- It produces cluster of orange coloured tubular flowers.
- The parasite produce dark red coloured berries.
- The birds are attracted by the fruits and act as dissemination agents.

» The seeds dropped on the crevices of the stem germinate and produce haustoria to attach to the host and proliferate in large numbers.

Giant-mistle toe: *Dendrophthoe falcata*

Management

» Removal of parasite affected branches and apply Bordeaux paste at cut ends. Collect the seeds of parasite and destroy.

Selected References

Arun Kumar AN, Joshi G. 2012. Incidence of sandal spike symptoms in a one-year-old plantation in Karnataka. Curr. Sci. 103: 613–614.

Barber CA. 1903. Report on spike disease in sandalwood trees in Coorg. Indian For. 29: 22–31.

Nagaveni HC, Sundararaj R, Vijayalakshmi G. 2014. First report of canker disease on Indian sandalwood (*Santalum album Linn.*) in India. Journal on New Biological Reports 3(2):120–124.

Nayar R, Ananthapadmanabha HS, Venkatesh KR. 1980. Seedling diseases of Sandal. Ind. J. Forestry., 3(1): 24-25.

Remadevi OK, Nagaveni HC, Muthukrishnan R. 2005. Pests and diseases of sandalwood plants in nurseries and their management. Working Papers of the Finnish Forest Research Institute 11 Fifth meeting of IUFRO working party.S7.03.04. 69-76.

47 Diseases of *Shorea robusta* (Sal)

S. No.	Disease	Pathogen
1.	*Polyporus* root rot	*Polyporus shoreae*
2.	Heart rots	*Fomes caryophylli, Fomes fastuosus, Hymenochaete rubiginosa, Trametes cubensis, Trametes incertae*
3.	Stem canker and witches broom	*Macrophoma shoreae*
4.	Sooty moulds	*Meliola* sp., *Asterina* sp., *Capnodium* sp.
5.	Leaf spot	*Curvularia palliscens*

1. Polyporus root rot (patridge root rot and sapwood rot)

Symptoms

- » It is a sporadic disease in sal forests with high rainfall of India, especially north Bengal, Assam, Odisha, Bihar, Madhya Pradesh and Uttar Pradesh.
- » Infection is happened in both healthy and injured trees.
- » Root gets affected at distal end and infection progresses up to collar region.
- » Top drying of branches is observed which leads to death of trees.
- » Trees become wind-blown.
- » Sometimes epicormic branches develop.

- The fungus produces white pocket rot in the bark and sapwood and they decay.
- Small light yellow felt of mycelium of fungus is seen on bark of roots.
- The heartwood remains healthy.
- Root injury or damage leads to infection.
- Spreads by root contact and root grafting.
- Fire protection, high soil moisture and heavy weed growth increases the disease.

Fungus: *Polyporus shoreae*

Management

- Control burning reduces soil moisture and check weed growth that reduces diseases incidence.
- All dead, dry stumps and drying trees should be removed.

2. Heart rots

a. High sal forests:

- Heart rot is seen as butt and trunk infections.
- Though heart wood is protected by thick bark and sapwood, due to aging, wounds on trees serve an opening of entry for pathogen.
- Wounds are caused by fire injury, breaking of branches, animal browsing, lightening and frosts.
- Dead dried branches/ stubs invite infection.

b. Coppice forests

- Infection establishes on stools and progresses up to the tree when heart rot develops in coppice shoots.
- Wounds caused during coppicing invite infection.
- Heart wood of stool is first affected; later heart wood of coppice is affected.

Symptoms of heart wood decay

i. **Punk knots:** Soft punky tissues of decayed wood having cylindrical cores of dia up to 2 cm and up to 6 cm in length emerge horizontally through the bark and open on the surface of bole a as pox marks/eyes. Later it becomes hard and black.

ii. **Swollen boles:** Distinct swellings in the bole leads to bottle shaped appearance. Also hollow sound in bole when hit with axe.

iii. **Sporophores** are seen on dead branches.

iv. **External injuries** on the branches, tops, openings and scars in the bole, fire wounds, flow of resins, crown drying, forking of main stem, cracks and protruberances in the bark.

Heart wood causing fungi

a. *Fomes caryophylli* (Syn: *Phellinus caryophylli***)**

Infection reported in Sal, *Terminalia, Anoigeissus latifolia* and *Mallotus phillipipensis*

Symptoms: Typical punk knots, yellow mycelial mat on wood later wood stained as brown colour and turns to white rotted area in pockets

Fungal characters: Sporophores sessile, perennial or annual, small, woody, upper surface brown –black and hymenial surface dull brown with stratified pores.

b. *Fomes fastuosus*

Infection reported in coppice sal forests, broad leaved trees and *Terminalia* sp.

Symptoms: Causes infection through fire injury, leads to butt rot. Heart wood stained as brown colour and turns to white rotted area. In stumps, affects coppice growth.

Fungal characters: Sporophores sessile, perennial, hard and large. Upper surface dull brown, hymenial surface dark brown, pore tubes not stratified.

c. *Hymenochaete rubiginosa*

Infection reported in broad leaved trees, sal and oak

Symptoms: Heart wood stained as brown colour and turns to white rotted area. Grows on stumps of dead trees, on logs and converted timber.

Fungal characters: Sporophores sessile, resupinate, corky, rigid on drying, upper surface exhibit dark brown, concentrically zonate, and hymenial surface become brownish.

d. *Trametes cubensis*

Infection reported in sal and *Terminalia* sp.

Symptoms: Causes brown cuboidal rot, white mycelium develops in cracks of decayed wood.

Fungal characters: It is a poroid bracket fungus. Sporophore is sessile, applanate, corky, rigid on drying. Upper surface light brown, becomes black at base, hymenial surface light brown.

e. *Trametes incertae*

Infection reported in sal

Symptoms: Heart wood stained as dark brown colour and turns to white rotted area in pockets.

Fungal characters: Sporophores sessile, woody, had. Upper surface dirty white to grey, hymenial surface is woody brown.

Management

» Prevent frost injury by canopy adjustment.

» Control burning to be practiced.

» Fire control measures lower the disease incidence.

» Thinning to be done where dead and injured trees to be removed.

» Fellings: During felling operation, the lopping should not damage other trees.

» Pathological rotation: is the age at which the progress of decay is equal to the rate of formation of new wood. Severely affected stands may be worked based on the basis of pathological rotation age.

» Wound dressing: In case of any residual wound on trees, the trees can be dressed with chemicals.

» Coppice stands: Here, the stools are dressed with chemicals to avoid infection.

» Even aged trees: Since even aged trees are more prone to frost, snow injury, the wounds are more in these trees that may lead to decay by heart rot. So practice felling early in even aged trees.

3. Stem canker and witches broom

Symptoms

- » Knotted twigs, prominent cankers with terraced callus and formation of witches broom.
- » Infects smaller branches that is injured by hailstones.

Fungus: *Macrophoma shoreae*

4. Sooty moulds

Symptoms

- » Honey dew secretion from sucking pests invites sooty black growth on the lead that blocks sunlight and prevents photosynthetic activity

Fungi: *Meliola* sp., *Asterina* sp., *Capnodium* sp.

Management

- » Spray starch solution 5%

5. Leaf spot

Symptoms

- » The spots were small brownish and more or less circular.
- » The centre of the spot was brown to dark brown surrounded by yellow halo mostly on young leaves.
- » The spots enlarged with maturity of leaves and varied from one to many.
- » They were irregular in shape and scattered all over the leaf surface.
- » In severe infection, individual spots coalesce together and most part of the leaves dried.

Fungus: *Curvularia palliscens*

Management

- » Control burning reduces soil moisture and check weed growth that reduces diseases incidence.

» All dead, dry stumps and drying trees should be removed.

Other Diseases

6. **Thread blight:** *Marasmius gordipes*

7. **Leaf spot:** *Cercospora* sp.

8. **Top dying:** *Hypoxylan* sp.

9. **Root and stem rot:** *Xylaria* sp.

10. **Phanerogamic parasites:** *Dendrophthoe falcata*

Selected References

Bagchee K. 1959. Fungal diseases of sal (*Shorea robusta* Gaertn.), Part 5: the heart-rot of sal caused by Trametes incerta (Curry) Cooke. Forest Research Institute (Dehra Dūn, India)

Bagchee K. 1959. Fungal diseases of sal (*Shorea robusta* Gaertn.), Part 2: Secondary parasite of sal. Forest Research Institute (Dehra Dūn, India)

Bagchee K. 1962. Fungal diseases of sal (*Shorea robusta* Gaertn.), Part 4: the heart-rot of sal caused by *Fomes caryophylli* (Rac.) Bres. Forest Research Institute (Dehra Dūn, India)

Bakshi BK. Boyce JS. 1959. *Polyporus shoreae* root rot on Sal. Indian Forester 85, 656-658.

48 Diseases of *Syzygium cumini* (Java Plum)

S. No.	Disease	Pathogen
1.	Leaf spot and fruit rot	*Glomerella cingulata*
2.	*Phoma* leaf spot	*Phoma tropica*
3.	*Neopestalotiopsis* leaf spot	*Neopestalotiopsis clavispora*
4.	Leaf spot and blight	*Cylindrocladium quinqueseptatum*
5.	Leaf blight	*Pestalotiopsis* sp
6.	Tar spot	*Rhitisma pongamiae*

1. Leaf spot and fruit rot

Symptoms

- The disease affected leaves show small-scattered spots of light brown or reddish brown colour.
- Fruits affected with fruit rot disease rot and shrivel.

Fungus: *Glomerella cingulata*

Management

- Spray with Zineb 0.2% or Bordeaux mixture 4:4:50.

2. *Phoma* leaf spot

Symptoms

» The symptoms appeared as scattered reddish brown speck in abundant numbers on dorsal surface of leaf which later develop to small, circular to irregular reddish brown spots with distinct pale brown to ash coloured centre.

» Further the spots got coalesced to form irregular shaped enlarged blighting covering entire leaf lamina.

» Upon disease severity progression leaves showed drying and upward curling.

Fungus: *Phoma tropica*

Management

» Spray with Difenconazole 0.1% or Carbendazim 0.2% or Hexaconazole 0.1% or Propiconazole 0.1%.

3. *Neopestalotiopsis* leaf spot

Symptoms

» Leaf spot began as a black dot with a distinct yellow halo.

» These dots enlarged to a diameter of 3 to 5 mm and progressed to form deep brown, circular, elliptical or irregular lesions with a dark black halo around the margin.

» In a humid environment, black, sessile, and discoid conidiomata developed on the spots and exuded a pink spore mass that turned brown with age.

Fungus: *Neopestalotiopsis clavispora*

Management

» Spray with Carbendazim 0.2% or Propiconazole 0.1%.

4. Leaf spot and blight

Symptoms

» Symptoms like greyish water-soaked spots which coalesce and develop into extensive necrotic areas, usually along the leaf margins and tips.

- Under favourable conditions of high humidity and frequent rainfall, the spots further spread to cover the entire area of the leaf blade and inducing leaf blight symptoms.
- Affected leaves turn reddish brown and are prematurely shed.

Fungus: *Cylindrocladium quinqueseptatum*

Management

- Spray with Mancozeb 0.1% or Carbendazim 0.05%.

5. Leaf blight

Symptoms

- Infected leaves shows extended elliptical grey center surrounded by brown margin which coalesce and develop into extensive necrotic areas, usually along the leaf margins and tips.
- Affected leaves turn greyish brown and are prematurely shed.

Fungus: *Pestalotiopsis* sp

Management

- Spray with Thiophanate methyl 0.1% or Carbendazim 0.1%.

6. Tar spot

Symptoms

- Initially, the upper leaf surface and produces black cushion-like stromata which may occur scattered or in clusters.
- In severe stage, the defoliation and drying symptoms are seen on the other branches.

Fungus: *Rhitisma pongamiae*

Management

- Remove severely affected trees.
- Spray with Propiconazole 0.5% or *Trichoderma viride* 20g/ tree.

Other Diseases

7. **Die-back and canker**: *Cadophora luteo-olivacea, Diplodia sapinea, Diplodia seriata, Neoscytalidium hyalinum, Phaeoacremonium fraxinopennsylvanicum, Phaeoacremonium krajdenii, Phaeoacremonium parasiticum* and *Pleurostoma richardsiae*

8. **Truck decay**: *Cryptococcus gattii, Cryptococcus neoformans*

Selected References

Arushi Mehrotra, Mehrotra MD. 2000. Leaf spotting and blight, a new disease of *Syzygium cumini* by two *Cylindrocladium* species from India. Indian Journal of Forestry, 23:4,496-500.

Banerjee A, Rana D, 2020. Leaf spot disease of *Syzygium cumini* caused by *Neopestalotiopsis clavispora* newly reported in India. Plant Disease. https://doi.org/10.1094/PDIS-09-19-2018-PDN.

Bhanumathi A, Rai VR. 2007. Leaf blight of *Syzygium cumini* and its management in vitro. Australasian Plant Disease Notes 2, 117–121. https://doi.org/10.1071/DN07049

Nagarjun N, Suryanarayana V. 2016. Documentation, characterization and management of leaf spot of *Syzygium cumini* (L.) Skeels. J. Farm Sci., 29(3): 381-385.

Nataliya K, Suryanarayana V, Thippeshi LC, Naik ST. 2011. Evaluations of different fungitoxicants against *Rhitisma pongamiae* sp. nov causing tar spot of *Pongamia pinnata* (L.) Journal of Plant Disease Research. 6(1): 20-24.

49 Diseases of *Tamarindus indica* (Tamarind)

S. No.	Disease	Pathogen
1.	Powdery mildew	*Oidium* sp.
2.	Leaf spot	*Colletotrichum gloeosporioides*
3.	Leaf blight	*Alternaria alternata*
4.	White root rot and wood rot	*Ganoderma lucidum, Lenzites palisoti*

1. Powdery mildew

Symptoms

» The powdery mildew disease develops as a minute white patch on the upper surface or lower surface, which later extends to cover the entire leaf lamina.

» As the disease progresses the powdery growth turns to light brown patches leading to defoliation.

» The infection may also occur on the flower stalk and flowers resulting in the shedding of flowers and young fruits.

» The white powdery growth of the pathogen represents the mass of conidiophore and conidia.

» In the advanced stage of the disease the fungi produce the sexual fruiting body called Cleitothecia.

Fungus: *Oidium* sp.

Management

- Spray with Wettable sulphur 0.2%.

2. Leaf spot

Symptoms

- The infection starts as water soaked lesions, which later turns to minute black spot with a dark margin.
- As the disease progresses, the spots enlarge and cause irregular blighting of the leaf lamina.
- The blighted leaf tissues show a hundreds of minute black coloured fruiting body of the fungus in concentric rings.
- The type of fruiting body is acervulus with a large number of hyaline, oblong, single celled conidia which helps in the secondary spread of the disease.
- Sometimes infection also spread to the tip of the young twigs which started drying back and this is named as "die back" symptom, leading to complete drying of the young shoot.
- The extensive dieback symptom leading to blighting of the new shoots is called Twig blight.
- In addition, the fungus also cause infection on the fruits as a black depressed lesion with a dark margin which cause extensive rotting of the fruit skin and internal tissues, named as fruit rot or fruit spot.

Fungus: *Colletotrichum gloeosporioides*

Management

- Spray with Mancozeb 0.2% or Copper oxychloride 0.2% twice at 15 days interval.

3. Leaf blight

Symptoms

- Brown water soaked lesions develop on the leaf and spreads fast to occupy major portion of the leaf leading to complete drying of the lamina.

» The disease can easily identified by presence of concentric rings in the blighted area showing the sporulation of the fungus.

Fungus: *Alternaria alternata*

Management

» Spray with Mancozeb 0.2% or Copper oxychloride 0.2% twice at 15 days interval.

4. White root rot and wood rot

Symptoms

» The disease can be visualized as yellowing and dropping of leaves in one or two branches of the tress, subsequently followed by drying of the branches.

» As the disease progresses, the defoliation and drying symptoms are seen on the other branches.

» The fungus grows extensively from the bark and causes white rot of the sap wood which becomes spongy.

» The dried trees show characteristic fruiting bodies of the fungus which arise from the base of the trees at the soil level.

Fungi: *Ganoderma lucidum*, *Lenzites palisoti*

Management

» Remove severely affected trees.

» Root feeding is also adopted where the root region is exposed; actively growing young root is selected and given a slanting cut at the tip.

» The root is inserted into a polythene bag containing fungicidal solution containing 2g of Aureofungin soil and 1 g of copper sulphate in 100 ml of water.

Other Diseases

5. Leaf spots: *Bartalinia robillardoides*, *Exosporium tamarindi*, *Hendersonia tamarindi*, *Pestalotia poonensis*, *Phyllosticta tamarindicola*, *Phyllosticta tamarindina*, *Prathigada tamarindi*, *Xanthomonas tamarindi*, *Sphaceloma* spp., *Stigmina tamarindi* and *Cercospora tamarindi*

6. **Seedling blight**: *Macrophomina phaseolina* and *Rhizoctonia solani*

7. **Sooty mould**: *Meliola tamarindi*

8. **Stem infection**: *Fracchiaea indica*

9. **Stem rot**: *Pholiota gollani*

10. **Trunk and root rot**: *Stereum nitidulum*

11. **Collar rot**: *Phytophthora nicotianae* var. *nicotianae*

12. **Stem canker**: *Hypoxlon nectrioides*

13. **Bark parasite**: *Myriangium tamarindi*

14. **Stone fruit**: *Pestalotia macrotricha*

15. **Sap rot**: *Xylaria euglossa*

16. **Brownish sap rot**: *Polyporus calcuttensis*

17. **White rot**: *Trametes floccose*

18. **Black pod**: *Aspergillus niger*

Selected References

Brownie FG. 1968. Pests and Diseases of Forest Plantation Trees. Clarendon Press Oxford: 1330.

Lokesha S, Shetty HS. 1991. A *Pestalotia* species causing stony fruit disease in tamarind. International Journal of Tropical Plant Diseases, 9(2): 179-181.

Mukherji KG, Bashin J. 1986. Plant Diseases of India: A Source Book. New Delhi. Tata McGraw Hill. 468.

Siddaramaiah AL, Kulkarni S. 1982. Control of powdery mildew disease of tamarind seedlings. Indian Forester, 108(5): 361-364.

50 Diseases of *Tectonia grandis* (Teak)

S. No.	Disease	Pathogen
1.	Leaf rust	*Olivea tectonae*
2.	Powdery mildew	*Uncinula tectonae*
3.	*Colletotrichum* leaf spot	*Colletotrichum gloeosporioides*
4.	*Phomopsis* leaf spot (Top dying)	*Phomopsis tectonae*
5.	*Sclerotium* leaf spot	*Sclerotium rolfsii*
6.	*Pseudoepicoccum* zonate leaf spot	*Pseudoepicoccum tectonae*
7.	*Rhizoctonia* leaf blight	*Rhizoctonia solani*
8.	*Alternaria* leaf blight	*Alternaria alternata*
9.	Thread blight	*Marasmius* sp.
10.	Butt rot/ Root rot	*Polyporus zonalis, Helicobasidium compactum, Rigidoporus lignosus, Rigidoporus zonalis*
11.	*Peniophora* root rot	*Peniophora rhizomorpho-sulphurea*
12.	*Ganoderma* root disease	*Ganoderma lucidum, Lenzites stereoides, Hexagonia apiaria*
13.	Heart rot	*Perenniporia tephropora*
14.	Basal root rot	*Phellinus noxius*
15.	*Armillaria* root rot	*Armillaria heimii, Armillaria mellea, Armillaria tabescens*
16.	*Helicobasidium* root rot	*Helicobasidium compactum*

17.	*Fomes* root rot and die-back	*Fomes lamaoensis, Fomes lignosus, Fomes noxius*
18.	Pink disease	*Phanerochaete salmonicolor*
19.	Die-back	*Phialophora richardsiae*
20.	Stem rot	*Marasmiellus ignobilis*
21.	Stem canker	*Nectria haematococca*
22.	Bacterial rot	*Pseudomonas tectonae*
23.	Bacterial wilt/ Bacterial collar rot	*Ralstonia solanacearum*
24.	Mosaic	Virus
25.	The mistletoe	*Dendrophthoe falcata* var. *pubescens*

1. Leaf rust

Symptoms

» The rust occurs almost round the year, it is most prevalent during winter season.

» The upper surface of the leaf show dull green flecks corresponding to the orange yellow uredinia on lower surface.

» The pustules on the upper surface become necrotic and appear as brown spots.

» The upper leaf surface presents a grey appearance due to the formation of flecks, which correspond to the position of sori on the lower surface.

» Infected leaves fall off prematurely resulting in retardation of plant growth.

» The disease is common in nursery and young plantations.

Fungi: *Olivea tectonae* (Syn.: *Uredo tectonae*); *Aecidium effusum*

Management

» The infected seedlings can be segregated and kept in isolation.

» Severely infected and dead seedlings can be burnt away from the nursery to prevent the spread of the disease.

» Spray *Olivea* sp hyperparasites *Acremonium recifei* and *Cladosporium oxysporum*

» The disease may be controlled in the nursery by the application of Wettable sulphur 0.1% on both sides of the leaves.

2. Powdery mildew

Symptoms

» Powdery mildew of teak occurs in plantations of all age groups but mostly in plantations of age above 15 years.

» The powdery mildew disease develops as a minute white patch on the upper surface or lower surface, which later extends to cover the entire leaf lamina.

» As the disease progresses the powdery growth turns to light brown patches leading to defoliation.

» The infection may also occur on the flower stalk and flowers resulting in the shedding of flowers and young fruits.

» The white powdery growth of the pathogen represents the mass of conidiophore and conidia.

» In the advanced stage of the disease the fungi produce the sexual fruiting body called Cleistothecia.

» The metabolic changes in plants take place which lead to drying of infected leaves.

Fungi: *Uncinula tectonae, Phyllactinia guttata* and *Phyllactinia corylea*

Management

» Immediate removal of infected leaves helps to prevent the disease spread.

» Release of lady bird beetle minimizes the powdery mildew inoculum.

» Spray with Wettable sulphur 0.1% or Triademefon 0.2% or Quinomethionate 0.1%.

3. Colletotrichum leaf spot

Symptoms

» The leaf spot disease is of common occurrence in plantations.

» It usually appears on mature leaves during the monsoon (July/August).

» Trees of all age groups are susceptible to this disease.

- Irregular leaf spots, light to dark brown in colour, with a pale margin are usually found on mature leaves, which develop near the tip and along the margin of the leaves.
- The individual spots coalesce to form large irregular spots, which cause drying up of leaves and consequently premature defoliation.
- The disease spreads laterally in the nursery through overlapping foliage of the adjoining seedlings often resulting in group blighting of seedlings.
- In each case of severe infection, defoliation is high.

Fungus: *Colletotrichum gloeosporioides* state of *Glomerella cingulata*

Management

- Immediate removal of infected plants helps to prevent the disease spread.

4. Phomopsis leaf spot (Top dying)

Symptoms

- The damage potential is high in nurseries.
- Spots appear as minute dark brown dots, 2 - 3mm in diameter.
- Spots enlarge to 5 - 8 mm in diameter and turn to light pale brown with dark brown outline.
- The margin grows out wards forming one to three dark brown concentric rings around the light coloured centre spot.
- The leaf spots cause considerable damage to the photosynthetic area.
- Where more than 50 per cent of the area is covered with necrotic lesions, leaves are defoliated prematurely.
- Hence more serious impact in nurseries is observed.

Fungi: *Phomopsis tectonae* and *Phomopsis variosporum*

Management

- Immediate removal of infected plants helps to prevent the disease spread.

5. Sclerotium leaf spot

Symptoms

- » The disease was characterised by the initial dull-brown coloured small leaf spots, 2-3 mm across, which under prolonged wet periods increased rapidly in size irregularly, often covering more than one-fourth of the lamina.
- » Frequently when the midrib or petiole was attacked, leaves wilted and dried up; rarely, tender parts of the stem were also found to be infected and growing shoots were killed.
- » A profuse mycelial growth was generally observed on the lower surface of the necrotic area of leaf.
- » Abundant white coloured sclerotia were produced on the mycelial strands, becoming dark when mature and dispersed by rain drops and wind to lower leaves, where they caused fresh infections.

Fungus: *Sclerotium rolfsii*

Management

- » Immediate removal of infected plants helps to prevent the disease spread.

6. Pseudoepicoccum zonate leaf spot

Symptoms

- » The disease incidence was usually located in humid tracts plants (1- to 4-year-old) continued to affect the high in plantations.
- » Though trees of any age group were attacked young were more susceptible than older ones.
- » The leaf spot was characterised by several well demarcated dark brown concentric rings.
- » Initially minute dark brown spots, 2-3 mm across, developed on the leaves which under high humidity enlarged by adding brown rings, 1-2 mm wide.
- » Subsequently when more rings were added to the spot it gave an appearance, of the target, hence the name target leaf spot.

- Nearby spots coalesced to form large necrotic area on the leaf.
- Occasionally a single spot covered about half of the leaf lamina.
- Severely infected leaves were prematurely defoliated, during monsoon, light greyish brown mycelial growth was frequently observed at the margin of the spot on the abaxial side from which the pathogen could be isolated easily.

Fungus: *Pseudoepicoccum tectonae*

Management

- Immediate removal of infected plants helps to prevent the disease spread.

7. *Rhizoctonia* leaf blight

Symptoms

- The infected plants show water soaked grayish brown patches that enlarge rapidly and cover a large part or the entire lamina.
- The blighted leaves often show holes in the infected portion as a result of shedding of infected tissues during heavy rains.
- The infected leaves dry up and are eventually shed.
- The disease spreads laterally in the nursery through overlapping foliage of the adjoining seedlings often resulting in group blighting of seedlings.
- In each case of severe infection, defoliation is high.

Fungi: *Rhizoctonia solani*, *Thanatephorus cucumeris*

Management

- Immediate removal of infected plants in the nurseries helps to prevent the disease spread.
- Avoid water stagnation in the field.
- Soil drenching with VAM and *Trichoderma viride.*
- Soil drenching with mancozeb 0.1%.

8. *Alternaria* leaf blight

Symptoms

» Brown water soaked lesions develop on the leaf and spreads fast to occupy major portion of the leaf leading to complete drying of the lamina.

» The disease can easily identified by presence of concentric rings in the blighted area showing the sporulation of the fungus.

» The fungus produces brown coloured conidia with both vertical and horizontal septa with a beak at the top.

Fungus: *Alternaria alternata*

Management

» Immediate removal of infected leaves in the nurseries helps to prevent the disease spread.

» Spray with Carbendazim + Mancozeb 0.2% or Chlorothalanil 0.1%.

9. Thread blight

Symptoms

» In the affected leaves, on the lower surface, a white mycelium originating from a thick light brown cord following the main vein and petiole.

» Mycelial cords are penetrated on leaves and branches, the infection portion dying and turning brown, thin and papery.

Fungus: *Marasmius* sp.

Management

» Immediate removal of infected leaves and branches helps to prevent the disease spread.

» Spray with Copper oxychloride 0.25% or Benomyl 0.1%.

10. Butt rot/ Root rot/ Top dying

Symptoms

- The fungus attacks heart wood, sometimes soft wood mostly colonize on teak stumps after coppicing or thinning.
- Initially the fungus causes white mottled rot with yellow lines, later pockets develop in the bleached area.
- Decay of roots and butts result in wind throw of trees.

Fungi: *Polyporus zonalis*, *Helicobasidium compactum*, *Rigidoporus lignosus*, and *R. zonalis*

Management

- Immediate removal of infected leaves and branches helps to prevent the disease spread.

11. Peniophora root rot

Symptoms

- The fungus attacks only the sapwood in which it causes yellow fibrous rot.
- The fungus rapidly migrates from decayed stumps into the coppice shoots and roots and produce thick and rarely branched rhizomorphs.
- The rhizomorphs are initially white and deep yellow in later.
- The infection restricted in the roots and affected trees become wind-blown due to decayed roots losing foothold in the soil.

Fungus: *Peniophora rhizomorpho-sulphurea*

Management

- Avoid prolonged wetting of roots.
- In disease prone area coppicing of teak may be avoided.
- Prevent the injury of seedlings.
- Select healthy stumps for platings.

- Removal of basidiocarp in the trunk and swabbing with Bordeaux mixture 1% helps to prevent the spread of the disease.
- Spot drenching with Bordeaux paste 10%.

12. Ganoderma root disease

Symptoms

- The disease can be visualized as yellowing and dropping of leaves in one or two branches of the tress, subsequently followed by drying of the branches.
- As the disease progresses, the defoliation and drying symptoms are seen on the other branches.
- The fungus grows extensively from the bark and causes white rot of the sap wood which become spongy.
- The dried trees show characteristic fruiting bodies of the fungus which arise from the base of the trees at the soil level.

Fungus: *Ganoderma lucidum*

Other wood decay fungi: *Lenzites stereoides* and *Hexagonia apiaria.*

Management

- Removal of basidiocarp in the trunk and swabbing with Bordeaux mixture 1% helps to prevent the spread of the disease.
- Spot drenching with Bordeaux paste 10%.

13. Heart rot

Symptoms

- In teak, the decay of wood at the center of the trunk and branches.
- Fungi enter the tree through wounds in the bark and decay the heartwood.
- The diseased heartwood softens resulting in trees being structurally weaker and prone to breakage.
- A good indication of heart rot is the presence of mushrooms or fungus conks on the tree.

Fungi: *Perenniporia tephropora* and *Phellinus caryophylli*

Management

» Removal of basidiocarp in the trunk and swabbing with Bordeaux mixture 1% helps to prevent the spread of the disease.

» Spot drenching with Bordeaux paste 10%.

14. Basal root rot

Symptoms

» The infected trees exhibit as yellowing and dropping of leaves, subsequently drying of the branches.

» The infection as indicated by bark depression, rotting at the root collar and signs of the presence of woodborers on the rotted portions, such as insect frass, bore holes and larvae on the affected trees.

» These symptoms and signs could be found on the tree stem up to 40 cm from ground level.

» Several affected trees were wind thrown, as the anchoring root system had been severely decayed.

» Below ground symptoms include the presence of dark brown decayed zones and irregular golden-brown honey comb like pockets of hyphae in the roots.

» In the bark, diseased zones were distinguishable by the presence of dark brown discolored patches, which were soft and friable when crushed with fingers.

» In the wood, the diseased zones were indicated by pale brown discoloration but the wood structure remained tough even though it had been penetrated by the fungal hyphae.

» In the majority of the dead trees the fungus had penetrated the wood, rotting the root systems, and progressed upwards to girdle the root collar.

» In affected trees brownish to dark brown resupinate basidiomata of *P. noxious* were found, forming a sock-like wrapping around the root systems.

» This rhizomorph becomes encrusted with age, turning a dark brown colour with clay soil adhering to the root surface.

» The pathogen preferred to survive in more acidic soil.

Fungus: *Phellinus noxius* (Syn: *Fomes noxius*)

Management

» Removal of basidiocarp in the trunk and swabbing with Bordeaux mixture 1% helps to prevent the spread of the disease.

» Spot drenching with Bordeaux paste 10%.

***15. Armillaria* root rot**

Symptoms

» Wilting and yellowing of tree crowns as well as occurrence of white mycelia fans under the bark of infected roots and root collars of teak trees.

Fungi: *Armillaria heimii, Armillaria mellea, Armillaria tabescens.*

Management

» Spot drenching with Bordeaux paste 10%.

***16. Helicobasidium* root rot**

Symptoms

» In the older plots the larger roots had proliferated from just above the highest level of disease attack.

» Initially dark rusty brown velvety band of fungal mycelium developed at the collar butt region of affected trees.

» Later on typically die-back symptoms appeared in these trees and the disease spread in larger areas in patches.

» The disease appeared to have been checked and confined to the lower portion, where it was no longer active; and colonization of the affected portion had been taken over by other soil saprophytes.

» Most of the diseased trees were associated with the cinnamon brown to purple, felty, fungal mats, of which several were bearing fertile hymenial patches.

- In severe, decline of the foliage, production of sprouts followed by death of the plants.

Fungus: *Helicobasidium compactum*

Management

- Spot drenching with Bordeaux paste 10%.

***17. Fomes* root rot and die back**

Symptoms

- The noticeable symptoms of withering teaks were yellowing, wilting and foliaceous organ desiccation.
- At the first stage of the disease, the parasite entered through wounds around the neck or the roots.
- Attacked tissues underwent deep decay.
- The decaying of woody tissues increased along the axis of the rods to the edge of the plants. When the plants were young and less lignified, the degradation invaded the trunk, and the tree decayed faster.
- On lignified trees, rot extended laterally on the same side of the tree trunk, along the affected root.
- Trunk contamination of adult teaks resulted in the appearance of small dark dimpled areas that evolved into a vertical and unilateral necrotic band.
- The disease revealed itself on the pivot, deep into the ground, and moved into the strain of the teak.
- In the case of regeneration section, the infection reached the strains.
- The consequences of the pathologic attacks spread up throughout the plant and led to the withering of the tree.
- At this stage, certain trees carried the germs, the fruiting bodies (carpophores) of Fomes sp.

Fungi: *Fomes lamaoensis, Fomes lignosus, Fomes noxius.*

Management

- » Spot drenching with Bordeaux paste 10%.

18. Pink disease

Symptoms

- » Pink disease of teak was usually observed in young plantations (1- to 5-year old).
- » The disease was characterised by a pink encrustation over a canker, formed at the site of infection on the stem.
- » The infection occurred at any place on the main stem, which resulted in killing of inner bark comprising of phloem and cambial tissue and subsequent development of a canker.
- » When the stem was completely girdled by the invading mycelium, the portion of shoot above the canker was killed outright.
- » The bark at the canker showed longitudinal splitting during the dry weather.
- » Just below the canker epicormic shoots developed, one of which usually became the leader shoot in the following season.

Fungus: *Phanerochaete salmonicolor*

Management

- » Diseased tree can be managed by following proper hygiene practices.
- » Spray with Carbendazim 0.1% at a weekly interval.

19. Die-back

Symptoms

- » *Cossus cadambae* was found to be responsible for the entry of the fungus in the wood.
- » Initial insect attack at the base of the branch near the main stem.
- » The lower side branches were killed.
- » Besides decay in the wood, partial girdling of the inner bark due to tunnelling was possibly also responsible for the death.

- Usually dead branches snapped and fell off from the area of intense insect attack, which could be seen in the form of a sieve of large number of holes, more or less in a circular fashion.
- In due course of time, such areas characterised by a group of holes, were found all over the stem, except at the lower region, one to two meters above ground.
- But as the insect attack increased, gradually all the branches showed die-back and later the whole tree succumbed.
- The death of the tree, which possibly may take between 10-15 years depending upon the age at which the infection started and intensity of insect attack, was primarily due to the fungal decay affecting sap as well as heartwood.
- Infection in the wood was initiated from the larval galleries or tunnels, sides of which appeared dirty dark brown in colour.
- The browning extended laterally as well as up and downwards in the wood; it even spread in the lower portion of the trunk, where no insect attack was found.
- The whole trunk was girdled due to decay and insect tunnelling, the tree died.

Fungus: *Phialophora richardsiae*

Management

- Diseased tree can be managed by following proper hygiene practices.

20. Stem rot

Symptoms

- Generally the infection occurred on one side of the trunk near the ground.
- The affected area became soft and spongy with profuse mycelial growth.
- During the rainy season numerous white to pale cream coloured basidiocarps (fructifications.) developed on the affected area.

Fungus: *Marasmiellus ignobilis*

Management

- Diseased trunk can be removed and follow proper hygiene practices.

21. Stem canker

Symptoms

- Trees die from top downwards; leaves turn yellow and shed off prematurely.
- Water soaked lesions develop under the bark, followed by the appearance of conidiophores and conidia.
- Later, leaves are wilted and the formation of dark black bumps on the surface of the stem.
- These bumps erupt and fissured, elongated wounds develop into holes in the trunk.
- Also, produces stem cracking and bark necrosis in infected trees.

Fungus: *Nectria haematococca* imperfect state of *Fusarium solani*

Management

- Diseased trunk can be removed and follow proper hygiene practices.
- Collatral host plants such as *Lantana* sp. should be cleared from land near teak plantations.
- Space the trees to allow adequate light exposure and good air circulation.
- If infection occurs, eradicate the disease by scraping the wounded stem and then smear it with Copper oxy chloride.

22. Bacterial rot

Symptoms

- The infected nursery exhibit cotyledon rot, collar rot and wilt of seedlings in conventional seedbed nurseries as well as root trainer nurseries.
- Also, vascular wilt of plants in young plantations.
- Severe disease incidence resulted in partial to complete failure of nurseries and large-scale mortality of plants in young plantations.

Bacterium: *Pseudomonas tectonae*

Management

- Diseases in nursery can be managed by following proper hygiene nursery practices
- Spray of bactericides (Pantamycin/ Oxytetracycline/ Streptocycline at 1000 ppm as soil drench).
- Spread of disease from nursery to plantation can be managed by preparing planting material (stump/seedlings) from disease free nursery stock and following strict quarantine measures.

23. Bacterial wilt / Bacterial collar rot

Bacterial wilt

Symptoms

- Bacterial wilt occurs in young teak plantations varying from 6-month-old to 2-year-old.
- The disease manifests during warm and wet period, especially just after the onset of monsoon.
- The infection usually occurs through injury and occasionally through lenticels.
- The symptoms, characteristic of vascular wilt disease, were expressed in the following dry period.
- The bacterium causes systemic infection of vessels, which show necrosis and discolouration.
- Initially the bottom leaves of the affected plants turned yellow and showed scorching and browning of leaf tissues in between the veins.
- The upper leaves, near the apex became flaccid and drooped.
- Such plants wilted and failed to survive.
- The roots of the affected plants got discoloured and decayed.

Bacterial collar rot

Symptoms

- Bacterial collar rot disease was recorded in 5-month-old teak seedlings

» In the affected seedlings, initially the collar area just above the ground showed a slight shrinking. At this stage the top leaves became flaccid and drooped.

» The affected area turned blackish-brown and got further constricted, the seedlings showed scorching of leaves with pronounced symptoms of wilting.

» The wilted seedlings died soon.

» The freshly cut stem above the affected area was immersed in a clean glass of water, streaks of off-white bacterial ooze from the vascular elements confirmed the systemic bacterial infection.

Bacterium: *Ralstonia solanacearum* (Syn: *Pseudomonas solanacearum*)

» The bacterium was gram negative, non-spore forming, aerobic, motile with single polar flagellum, and rod-shaped with round ends. The bacterium occurs in the soil and infects the plants through injury. The pathogen plugged in the xylem and phloem tissues, which exhibits the vascular discoloration and tissue necrosis.

Management

» Provide proper sanitation methods.

» The affected plants should be uprooted carefully and burnt.

» Planting in water-logged areas and low-lying areas should be avoided.

» Seedbeds should be properly levelled.

» In disease prone nurseries, the soil should be sterilized with polythene mulching or formalin about 2 weeks before planting.

» Clayey soil should be avoided.

» Weeding operations should be carried out carefully so as not to cause any injury to seedlings.

» As the disease is easily manifested through an injury to the root system, utmost care should be taken during weeding and soil working operations.

» For casualty replacement, the planting should be done in a separate pit dug away from the site of the infected plant.

- The soil at the site of the infected plant should be drenched with Plantamycin 0.01% or Streptomycin sulphate, dazomet, oksonilik acid.

24. Mosaic

Symptoms

- The affected plants were easily distinguishable in the nurseries or plantations due to their mottled leaves.
- The affected leaves had a number of small to large irregular creamy white to light pale irregular spots, which gradually coalesced to form large chlorotic areas, characteristic of mosaic disease.
- Mosaic symptoms were also seen in very young leaves.
- Often only a few of the leaves of a plant either had only half of the lamina with mosaic symptoms and rest with normal colour or only some of the leaves in a seedling showed mosaic symptoms while the others were normal.
- No crinkling of leaves and stunting of plants were observed.

25. Phanerogamic parasites (The mistletoe)

Symptoms

- Initial infection was found to start at the lower part of branches in the middle regions of a host tree.
- The host tree branches infected by *D. falcata* shows gradual reduction in the growth and diameter compared to non-infected branches
- Reduce fruit yielding and timber yielding plants.

Parasite: *Dendrophthoe falcata* var. *pubescens* (Syn: *Loranthus longiflorus*)

Management

- Provide proper sanitation methods.
- The affected branches should be removed carefully.
- The tree at the site of the infected plant should be drenched with Bordeaux mixture 1%.

» Trunk injection with metribuzin.

» Spray paraquat 1% or 2,4-D 0.2%.

Other Diseases

26. **Wound infection**: *Auricularia polytricha*

27. **Leaf spot**: *Cercospora tectonae, Cercospora tectonigena, Cercospora apii, Pseudocercospora tectonae, Mycosphaerella tectonae, Phyllosticta tectonae* (*Asteromella tectonae*), *Spaceloma tectonae, Pseudofusicoccum adansoniae.*

28. **Brown leaf spot**: *Alternaria alternata* (Syn: *Torula alternata*)

29. **Leaf blight**: *Phoma glomerata, Phoma eupyrena, Phomopsis variosporum.*

30. **Leaf blotch**: *Acremonium tectonae*

31. **Foliage rot**: *Corynespora* sp., *Calderiomyces* sp.

32. **Dieback**: *Stemphylum* sp., *Diaporthe neoraonikayaporum, Diaporthe tectonae, Diaporthe tectonigena, Lasiodiplodia pseudotheobromae.*

33. **Stem wilt**: *Phaeoacremonium italicum, Phaeoacremonium tectonae*

34. **Canker**: *Corallomycetella elegans*

35. **Sapwood infection**: *Lyctus* sp., *Kretzschmaria deusta,* (Syn: *Ustulina deusta*)

36. **Wood rot fungi**: *Marasmiellus ignobilis*

37. **Root infection**: *Fusarium oxysporum, Plemotus commisciloilis, Xylaria thwaitesii*

38. **Root rot**: *Spongipellis spumeus*

39. **Heart rots**: *Bjerkandera adusta, Flavodon flavus, Ganoderma applanatum, Ganoderma curtisii, Ganoderma rivelosum, Ganoderma tornatum, Ganoderma australe, Ganoderma colossus, Phellinus lamaoensis, Polyporus rubidus, Polyporus shoreae, Polyporus zonalis, Thaxteriellopsis lignicola, Tubeufia tectonae*

40. **Die-back**: *Pleurostomophora richardsiae*

41. **Bacterial leaf spot**: *Xanthomonas melhusi*

42. **Algal parasite**: *Cephaleuros* sp. and *Stomatochroon* sp,

43. **Mistletoes**: *Loranthus* spp., *Macrosolen cochinchinensis, Phoradendron piperoides, Phthirusa adunca*

Selected References

Bakshi BK. 1976, Forest Pathology, Principles and Practices in Forestry. Controller of Publication, New Delhi, India, p.400.

Farid AM, Lee SS, Rosli HM, Maziah Z, Norwati, M. 2005. Incidence of teak basal root rot caused by *Phellinus noxius* in Malaysia. Australasian Plant Pathology 34: 277. https://doi.org/10.1071/AP05008.

Hocking D. 1967. Root Rot of Teak, Pest Articles and News Summaries. Section B. Plant Disease Control, 13:3, 276-281, DOI: 10.1080/05331846709432255.

Maheswarappa V. 2001, Studies on rust of teak caused by *Olivea tectonae* (Racib). M.Sc. Thesis., Univ. Agric. Sci., Dharwad (India).

Mahmudsudin Lee S, Ahmad Hj. Harun, 1993, A survey of heart rot in some plantations of *Acacia mangium* in Sabah. J Trop. For. Sci., 6: 37-47.

Mohanan C. 2009. Bacterial diseases of teak (*Tectona grandis* L.f.) in forest nurseries and plantations in Kerala and their management. Indian Journal of Forestry, 32 (1) 131-136.

Morrison DJ, Williams RE, Whitney RD. 1991. Infection, disease development, diagnosis and detection. In: Armillaria Root Rot Disease. Eds. Shaw, C. J., Kile, G. A. Agriculture Handbook No. 691. Washington, DC: Forest Service, United States Department of Agriculture, pp. 62-75.

Prasanth RH, Naik ST. 2010. Diversity of fungi in different teak ecosystems. Karnataka J. Agric. Sci., 23(2): 394-396.

Sharma JK, Mohanan C, Florence EJ. 1984, Diseases survey in nurseries and plantations of forest tree species grown in Kerala For. Res. Inst. Peechi, p.36.

Stack RW, Lamey HA. "Deciduous Tree Diseases." NDSU Extension Service and ND Agricultural Experiment Station.

Streets RB. 1982. The Diagnosis of Plant Diseases: a field and laboratory manual emphasizing the most practical methods for rapid identification. The University of Arizona Press: Tucson, Arizona.

Weaver PL. 1993. *Tectonia grandis* L.f. Teak. SO-ITF-SM-64.

51 Diseases of *Terminalia* sp. (Arjun)

S. No.	Disease	Pathogen
1.	Leaf spot	*Pestalotiopsis palmarum*
2.	Black nodal girdling	*Uredo* spp.
3.	Rust	*Sphaceloma terminaliae*
4.	Powdery mildew	*Phyllactinia terminaliae*

1. Leaf spot

Symptoms

» The disease appears on any part of the blade as circular to irregular patches, sometimes delimited by veins.

» The infected leaves show characteristic copper brown spots, just above the fungal growth, on the upper surface.

» Usually these are 2-8 mm in diameter but sometimes more extensive.

» Affected leaves become brittle.

» The necrotic area becomes dry, shrunken and severed partly or entirely from the leaf as a hole.

» Severely infected leaves from which much of the dead tissue has fallen away present an extremely ragged or insect eaten appearance.

- Leaf yield loss due to leaf spot disease has been reported to be 8-12%.
- Leaf spot occurs during July to November all over India when high humidity and high temperature prevail in the atmosphere.
- The disease incidence observed in both *T. arjuna* and *T. tomentosa*

Fungus: *Pestalotiopsis palmarum*

Management

- Proper cultural operations and field sanitation.
- Plucking and burning of diseased leaves.
- Spray with Carbendazim at 0.1% or Blue copper 0.5% or Mancozeb 0.5%.

2. Black nodal girdling

Symptoms

- Disease symptoms begin from nectar gland and gradually spread to lower petiole region, in due course of time petiole and internodal regions get covered with black sooty spores along with mycelia, forming girdle like appearance.
- Thus, the entire leaf becomes unsuitable for silkworm rearing.
- In both *T. arjuna* and *T. tomentosa* plants, about 7-10% crop loss is reported.
- The disease occurs during rainy and winter seasons.
- This disease also appears during July at most of the places, increases up to January and then declines.

Fungus: *Uredo* spp

Management

- Proper cultural operations and field sanitation.
- Plucking and burning of infected leaves.
- Spray with Carbendazim at 0.1%.

3. Rust

Symptoms

- Infected plants showed rust like symptoms on host plant.
- It causes small spots on leaves.
- Fruiting body of rust presented on lower surface of leaves.
- Rust produce white creamy color pustules on lower surface of leaves.
- Telutospores of rust were found round shaped, whitish and one celled.

Fungus: *Sphaceloma terminaliae*

Management

- Proper cultural operations and field sanitation.
- Plucking and burning of infected leaves.
- Spray with Carboxin at 0.1%.

4. Powdery mildew

Symptoms

- As the name indicates white powder like patches appear on the ventral surface of leaves, afterwards appearance of distinct dot like structures is noticed on the powdery patches.
- Under advanced stage, the disease shows chlorotic symptoms on dorsal surface of leaves with curved lamina.
- Affected leaves become yellow and fall down prematurely.
- The crop loss has been reported to be 25-30% in *T. arjuna* and 8-10% in *T. tomentosa*, though qualitative loss may be much more.

Fungus: *Phyllactinia terminaliae*

Management

- Proper cultural operations and field sanitation.

- Plucking and burning of diseases leaves as and when they appear.
- Spray with Carbendazim at 0.1% or Wettable sulphur at 0.2%.

Selected References

Bakshi BK. 1976, Forest Pathology, Principles and Practices in Forestry. Controller of Publication, New Delhi, India, p.400.

Butler EJ. 1997. Fungi of India, Biotech book Publication Delhi, page-220.

Pawar VP, Vidya A, Patil. 2011. Occurrence of powdery mildew on some wild plants from khandesh region of maharashtra state. Recent Research in Science and Technology 3: 94-95.

52 Diseases of *Ziziphus mauritiana* (Ber)

S. No.	Disease	Pathogen
1.	Powdery mildew	*Oidium erysiphoides* f. sp. *zizyphi, Pseudoidium ziziphi*
2.	Black leaf spot	*Isariopsis indica* var. *zizyphi*
3.	Rust	*Phakopsora zizyphi-vulgaris, Uredo ziziphi*
4.	Die-back	*Phytophthora* sp., *Fusarium solani*
5.	*Cladosporium* leaf spot	*Cladosporium zizyphi*
6.	*Alternaria* leaf spot	*Pleospora infectoria, Pleospora caricola, Pleospora passeriniana, Alternaria hartarum, Alternaria passeriniana, Alternaria chartarum, Annellophorella* sp
7.	*Cercospora* leaf spot	*Cercospora ziziphae, Cercospora jujubae*
8.	Anthracnose	*Colletotrichum gloeosporioides*
9.	*Phoma* fruit rot	*Phoma hissarensis*
10.	*Alternaria* fruit rot	*Alternaria alternata*
11.	*Trichothecium* fruit rot	*Trichothecium roseum*
12.	*Cladosporium* fruit rot	*Cladosporium hebarum*
13.	Red rust	*Cephaleuros virescens*
14.	Witches broom	*Candidatus* Phytoplasma
15.	*Septoria* leaf spot	*Septoria capensis*
16.	*Pestalotia* leaf spot	*Pestalotia subinae*

1. Powdery mildew

Symptoms

- White powdery growth appears on young leaves and causes them to shrink and defoliate.
- Small, white powdery growth on young fruits enlarges, coalesce and finally turn brown to dark brown.
- In severe cases, the whole fruit surface covered with the powdery mass.
- The spots turn into light brown to dark brown discolouration.
- The infected area becomes slightly raised and rough leads to premature dropping of fruits.
- Infected fruits become corky, cracked, mis-shapened and underdeveloped.
- Matured fruits turn rusty and unfit for marketing.

Fungi: *Oidium erysiphoides* f. sp. *zizyphi, Pseudoidium ziziphi*

Management

- Collect and destroy infected fallen leaves and fruits.
- Grow resistant varieties like Safeda Rohtak
- Grow tolerant varieties like Illaichi Jhajjar, Gola, Seb, Safed Rohtak and Mehrun.
- Spray Wettable sulphur 0.3% or Carbendazim 0.1% or Thiophenate Methyl 0.1% should be done during the fruit attains pea size.

2. Black leaf spot/ Black mould

Symptoms

- Sooty tuft like circular to irregular black spot develop on leaf surface.
- In severe infection, it covers a large area on the lower surface of the leaves and corresponding upper surface shows yellowish to brownish discoloration and finally drop off.

Fungus: *Isariopsis indica* var. *zizyphi*

Management

- » Grow resistant varieties like Seo-Bahadurgarhia, ZG-3, Safed Rohtak, Sanaur-1, Mundia, Banarsi Karaka, Banarsi Pewandi and Bagwadi
- » Spray with mycoparasite, *Hansfordia pulvinata* during early infection stage.
- » Spray with Mancozeb 0.2% or Zineb 0.2% or Wettable sulphur 0.3% or Copper oxy chloride 0.2% or Bordeaux mixture 2:2:50 or Captafol 0.05% or Carbendazim 0.1% or Thiophanate methyl 0.1%.

3. Rust

Symptoms

- » Small, irregular, reddish brown uredopustules appear on the lower leaf surface and later they cover the whole area of leaves.
- » The infected leaves dry and defoliate.

Fungus: *Phakopsora zizyphi-vulgaris, Uredo ziziphi*

Management

- » Grow resistant varieties like Banarsi, Seo, Katha Gurgaon, Gola Gurgaon-2, Dandan, Sanaur-1, Safeda Selected and Sanaur-3
- » Spray with Mancozeb 0.2% or Zineb 0.2% or Wettable sulphur 0.3% or Copper oxy chloride 0.2%.
- » Spray with Thiophanate methyl 0.1% or Propiconazole 0.75% or Mancozeb 0.25%.

4. Die-back

Symptoms

- » The pathogen causing dieback, tip dieback, graft union blight, twig blight, seedling rot, wood stain, stem-end rot, black root rot, fruit rot, dry rot, brown rot of panicle etc.
- » It is characterized by drying back of twigs from top downwards, particularly in older trees followed by drying of leaves which gives an appearance of fire scorch.
- » Internal browning in wood tissue is observed when it is slit open along with the long axis.

» Cracks appear on branches and gum exudes before they die out.

Oomycete: *Phytophthora* sp.

Fungus: *Fusarium solani*

Management

» Spray with Mancozeb 0.2% or Copper oxy chloride 0.2%.

5. *Cladosporium* leaf spot

Symptoms

» The symptoms of the diseases are produced on leaves in the form of small, light brown to brown irregular spots on the lower surface of leaves.

» Individual spots may coalesce to form large irregular areas on the leaf blade.

» The disease starts on leaves closest to the soil surface, where the fungus occurs.

Fungus: *Cladosporium zizyphi*

Management

» Grow resistant varieties like Sandhura Narnaul, ZG-3, Banarsi, Govindgarh Selection-3, Jhajjar Selection and Jogia

» Spray with Mancozeb 0.2% or Zineb 0.2% or Copper oxy chloride 0.2%.

6. *Alternaria* leaf spot

Symptoms

» This disease appears with the formation of small irregular brown spots on the upper surface of leaves and grey to brown spots on the lower surface.

» Many spots coalesce to form large patches and such leaves drop off.

Fungi: *Alternaria* anamorph of *Lewia (Pleospora) infectoria,*

Alternaria anamorph of *Pleospora caricola, Pleospora passeriniana,*

Alternaria hartarum, Alternaria passeriniana, Alternaria chartarum, Annellophorella sp.

Management

» Grow resistant varieties like Govindgarh Special, Gola, Gurgaon, Popular Gola, Seo-Bahadirgarhia, ZG-3, Sofed Rohtak and Jhajjar Special.

» Spray with Zineb 0.1% or Mancozeb 0.3% or Captafol 0.2% or Copper oxychloride 0.3%.

7. *Cercospora* leaf spot

» The disease manifests itself in the form of circular to oval spots, measuring up to 4 mm in diameter, epiphyllous, and yellow at first and turning brown surrounded by a dark brown margin.

» The spots grow larger and become visible on both sides of the leaves.

» The infected leaves fall off.

» *Cercospora* leaf spot is also found on Chinese jujube, causing leaf yellowing.

Fungi: *Cercospora ziziphae*, *Cercospora jujubae*

Management

» Grow resistant varieties like Safed Rohtak, ZG-3, Seo-Bahadurgarhia, Popular Gola, Rashmi and Jhajjar

» Spray with Mancozeb 0.2% or Zineb 0.1%.

8. Anthracnose

Symptoms

» Symptom of the disease on leaf was recorded as appearance of irregular, rough or corky reddish brown spots with yellowish margin, having the size of 2 to 3 mm in diameter.

» The spots generally appear on the upper surface of leaves and are not restricted by veins.

» Initially these spots were isolated but at later stage they may coalesced to form large patches on the infected leaves.

» On the fruit surface symptoms were small, circular to roughly circular, brown to black colored spots with depressed sporulating zone at the center.

» Size of spots quickly enlarges during color change of fruit peel from green to yellow and on each fruit usually 3-4 spots were recorded.

Fungus: *Colletotrichum gloeosporioides*

Management

» Spray with Carbendazim 0.2% or Chlorothalanil 0.2%.

***9. Phoma* fruit rot**

Symptoms

» The infected fruits remain small and develop slightly depressed, dark brown spots near the stem ends.

» The lesion becomes irregular in shape 15-25 mm in diameter.

Fungus: *Phoma hissarensis*

Management

» Spray with Mancozeb 0.2% or Zineb 0.2%.

***10. Alternaria* fruit rot**

Symptoms

» Slightly depressed, brown to dark brown, circular lesions appear on the fruit.

» Sometimes concentric rings are also present on these spots.

» The smaller spots coalesce to form larger spots.

Fungus: *Alternaria alternata*

Management

» Spray with Mancozeb 0.2% or Thiophanate methyl 0.1% or Metiram + Pyroclostrobin 0.2%.

11. Trichothecium fruit rot

Symptoms

- The disease is observed during the spring in the form of pink spots on the fruits.
- The fungus can survive in the soil for a long time.
- Fruits touching the soil may become infected and develop symptoms.

Fungus: *Trichothecium roseum*

Management

- Spray with Mancozeb 0.2% or Zineb 0.1%.

12. Cladosporium fruit rot

Symptoms

- The disease appears near the time of fruit ripening. Injured fruits become infected.
- The symptoms of the disease start from the tip of the fruit forming light brown to dark brown spots.
- Later, a greenish fungal growth is also seen on these spots.

Fungus: *Cladosporium hebarum*

Management

- Grow resistant varieties like Banarsi, Villaity, ZG-3, Govindgarh Sel-3, Jhajjar Selection and Zogia
- Spray with Tebuconazole + Trifloxystrobin 0.1%.

13. Red rust

Symptoms

- Red rust is characterized on ber as reddish brown, raised lesions on the upper surface of leaves.
- Spots were circular to irregular in shape having 2 to 3 mm diameter and few spots (3 to 5) were observed on a leaf.

» Spots are generally isolated from each other during initial stage of infection; however, they may coalesce to each other during maturity.

Alga: *Cephaleuros virescens*

Management

» Spray with Zineb 0.1% or Copper oxychloride 0.3%.

14. Witches broom

Symptoms

» The affected trees become weak and produce small, yellow leaves typical of other phytoplasma diseases.

» The disease is transmitted to healthy seedlings by budding and grafting with diseased scions

Bacterium: *Candidatus* Phytoplasma

Management

» Spray with Oxytetracycline 1000 ppm or Tetracycline hydrochloride 1000 ppm.

***15. Septoria* leaf spot**

Symptoms

» The affected leaves shows dark brown spots measuring 2-8 mm in diameter, smaller spots coalesce to form bigger spots.

Fungus: *Septoria capensis*

Management

» Spray with Tebuconazole + Trifloxystrobin 0.1%.

***16. Pestalotia* leaf spot**

Symptoms

» The affected leaves show irregular dark brown spots measuring 1.5-4 mm.

» Black sub-erumpent acervuli appear on the upper surface of the leaves.

Fungus: *Pestalotia subinae*

Management

» Spray with Tebuconazole + Trifloxystrobin 0.1%.

Other Diseases

17. ***Tandonella* leaf spot**: *Tandonella zizyphi*

18. ***Sirosporium* leaf spot**: *Sirosporium carissae*

19. ***Botryodiplodia* leaf spot**: *Botryodiplodia theobromae*

20. ***Corynespora* leaf spot**: *Corynespora ziziphae*

21. ***Septoria* leaf spot**: *Septoria capensis*

22. ***Pestalotia* leaf spot**: *Pestalotia subinae*

23. **Foliar blight**: *Pseudocercospora zizyphus-jujubae, Pseudocercospora ziziphina, Pseudocercospora zizyphicola, Pseudocercospora rhamnaceicola*

24. **Fruit and leaf spot**: *Phaeoramularia rhamnaecearum*

25. **Twig blight**: *Phomopsis ziziphina, Pestalotia funera*

26. **Blight**: *Diplodia zizyphiae*

27. **Wilt**: *Fusarium equiseti*

28. **Root rot**: *Cytospora sacculus*

29. **Thick rotten disease**: *Botryosphaeria obtusa*

30. **Foliar infection**: *Elsinoe zizyphi, Fusarium decemcellulare, Hypozylon hypomiltum, Mitteriella zizyphina, Myriangium zizyphi, Othia zizyphusjujuba, Poria* sp., *Sclerotium rolfsii*

31. **Post-harvest decay**: *Cladosporium tenuissimum, Fusarium pollidoroseum, Pythium aphanidermatum, Phoma nebulosa, Rhizoctonia solani, Alternaria alternata, Phoma destructive, Fusicoccum* spp., *Phoma capitulum, Phomopsis ziziphina, Cladosporium oxysporum, Curvularia lunata, Colletotricum gloeosporioides, Ceratocystis dipose, Ceratocystis morean* and *Trichothecium roseum*

32. **Fruit shrink**: *Coniothyrium olivaceum, Alternaria alternata, Dothiorella gregaria, Alternaria tenuis* (*Alternaria alternata*).

Selected References

Arjunan G, Parthasarathy S, Dinakaran D. 2019. Diseases of Horticultural Crops, Devi Publications, India.

Bhansali RR, Arya HC. 1978. A note on the occurrence of powdery mildew of ber (Zizyphus spp.) in Indian arid zone. Annals of Arid Zone, 17: 323- 325.

Chauksay P, Rai AN, Verma N. 2011. Some additions to the foliicolous fungi from the forest flora of Madhya Pradesh, India. In Microbial Biotechnology and Ecology, pp. 268–274. Eds D. Vyas, G.S. Paliwal, P.K. Khare and R.K. Gupta. New Delhi, India: Daya Publishing House.

Du Q, Zhao SF, Wu CL, Kong L, Zhang P. 2013. Root rot of Chinese Jujube (*Ziziphus jujuba*) caused by *Cytospora sacculus* in China. Plant Disease, 97, 1661

Gupta PC, Madaan RL. 1977. Diseases of fruits from Haryana – a new leaf spot disease of *Zizyphus mauritiana* Lamk. Current Science, 47, 237–238.

Gupta D, Razdan VK. 2010. First report of *Pestalotia funera* causing leaf spot disease of Ziziphus mauritiana from India.

Hsieh WH, Goh TK. 1990. *Cercospora* and Similar Fungi from Taiwan. Taipei, Taiwan: Maw Chang Book Company. 376 pp.

Jain SL, Rai AN, Mehta P. 2002. Additions to the genus *Corynespora* from India. Indian Phytopathology, 55, 51–56.

Jayarajan R, Cheema SS. 1972. Screening of ber varieties for resistant to leaf spot and powdery mildew diseases. Indian J. Hort., 29: 353.

Kapur SP, Cheema SP, Singh MP. 1975. Occurrence and control of powdery mildew of ber. (Zizyphus mauritiana Lam.) in Punjab. 1. Res., 12 : 26-29.

Kumar A, Kumar P, Singh HK, Saxena RP. 2003. Ber germplasm screening and management of black leaf sport disease under Eastern U.P. conditions. Journal of Applied Horticulture, 6, 43–44.

Sharma S, Kaushik RA, Panwar RD. 2003. Twig blight of ber caused by *Phomopsis zizipina* – a new report. Indian Phytopathology, 56, 236.

Tripathi SN, Rai AN, Verma NK. 2012. Some additions to the hyphomycetes from central India. Journal of Mycology and Plant Pathology, 42, 201–206.

Zhang ZY, Zhang H, Li TF. 2000. Taxonomy of the genus *Cladosporium* in China XXII. *C. platycodonis* sp. nov. and two new records. Mycosystema, 19, 308–311.

53 Wood Decay and their Management

The importance of Wood Decaying Fungi (WDF) was recognized only after the establishment of the concept of 'Heart rot' by Robert Hartig. Wood decaying fungi belonging mainly to 90 % of Basidiomycetes and 10% of Ascomycetes inhabit the live, dead or felled tree trunks (Logs), roots and coppiced wood. These fungi play a significant role in lignin decomposition in the forest and in turn, the organic matter is added to the soil and thereby carbon is returned to the atmosphere in the form of carbon dioxide. But their decay in commercially important trees in plantations leads to the destruction of valuable raw wood material. Moreover, these fungi also attack wood in service that leads to a significant loss. The majority of the wood-rot fungi belonging to the family Polyporaceae (bracket fungi), which includes the genus such as *Fomes*, *Polyporus* and *Ganoderma* causing severe losses of timber throughout the world. Wood Decaying Fungi affect the physical characters of wood such as colour, odour, strength and a specific gravity of wood.

I. Types of Decay Based on Position of Root

Wood decay fungi are separated into top rots and root or butt rots depending on which part of the tree is affected. Important agents of decay in trees are listed here

a. **Root decay fungi**: *Ganoderma resinaceum, Inonotus dryadeus, Meripilus giganteus, Armillaria* spp., *Ganoderma lipsiense, Grifola frondosa, Heterobasidion annosum, Kretzschmaria deusta, Pholiota squarrosa, Phaeolus schweinitzii, Phellinus weirii, Sparassis crispa* etc.

b. **Stem decay fungi**: *Daedaleopsis confragosa, Fistulina hepatica, Fomes fomentarius,*

Fomitopsis pinicola, Ganoderma adspersum, Inonotus hispidus, Inonotus radiates, Laetiporus sulphureus, Phellinus igniarius, Phellinus robustus, Piptoporus betulinus, Pleurotus ostreatus, Stereum sanguinolentum etc.

II. Types of Decay Based on Organisms

The breaking down of the wood tissues by microorganisms is termed as Decay. Decay or rot is mainly caused by fungi. However, Bacteria are also involved in decay but the process is slower when compared to the fungal invasion. Based on the nature of pathogens, the decay is classified into two types

a. **Bacterial Decay/Rot:** The decay caused by bacteria is a very slow and prolonged process. The bacteria colonize the ray parenchyma cells of sap wood and grow. Later, they attack the walls of the ray cells by producing some pectic enzymes leading to soft rots.

b. **Fungal Decay/ Rot:** Based on the nature of the destruction of the wood tissues, the wood-destroying fungi can be broadly grouped into three broad classifications as Soft rots, White rots and Brown rots.

i. **Soft rots:**

» Attacks wood in service.

» The fungi which cause soft rots are primarily members of Ascomycotina and Deuteromycotina (*Chaetomium, Ceratocystis*).

» When the wood is exposed to highly humid/ moist weather, soft rot occurs superficially on the surface of the wood due to the breakdown of the cellulose.

» It is most prevalent in the wood exposed to excessive moisture, such as in cooling tower slats, pilings, boats, fence posts, telephone poles, ponded logs and unpainted building siding.

» The organisms mainly attack polysaccharide constituents of secondary cell wall, while lignin is left relatively left unaffected.

» Rotting results in significant strength loss.

ii. **White rots**:

» The fungus is often associated with hard woods and decomposes all components of wood, including lignin.

- Mostly fungi belonging to Polyporaceae come under this group. Eg: *Ganoderma lucidum, Coriolus versicolor* (*Trametes versicolor*)
- Involves enzymatic delignification of lignin and cellulose, where the affected wood is reduced to spongy mass, to white pockets or streaks of various size separated by firm, strong wood, giving a bleached appearance to the wood.
- This type of rot is very common in hard woods when compared to soft woods
- The decomposed wood is usually white, but sometimes it may be yellow, tan or even light brown in colour.
- Dark lines (Zone lines) are seen in the decayed wood due to strings of tightly woven hyphae.
- White rotted wood exhibits only slight reduction in strength; however no major reduction in the pulp yields.

iii. Brown rots:

- Decays mostly soft woods only.
- The fungus involves initially a nonenzymatic depolymerization of cellulose and its associated pentosans involving superoxide radicals, iron and hydrogen peroxide decomposes the cellulose.
- The fungus removes cellulose and hemi cellulose in very short time from the wood leaving behind modified lignin as brown rot residues.
- The wood is reduced to a carbonaceous mass in various shades of brown, which can be powdered between fingers (Brown rot). Eg; Poria placenta, Polyporus sulphureus, Fomitopsis pinicola
- Brown rotted woods exhibit longitudinal swellings and shrinkage which gives rise to characteristic cubical checking pattern
- Mostly the cellulose and hemicelluloses enzymes are involved
- Brown rotted trees show considerable reduction in the strength
- Much reduction in the pulp yield is noticed

III. Types of Decay Based on Nature of Infection and Damage to the Wood

The damage to wood due to decay by fungi is classified as 'Staining' and 'Decay'.

1. **Staining:** Decay is associated with staining. Sometimes staining may occur without decay also. Staining occurs in sapwood like a dark coloured superficial moldy growth of fungi which can be easily removed off on the surface of the wood. These fungi do not have an enzyme system to break down the cell components but the fungal hyphae initially concentrate with in the ray cells and obtain nutrition from the cell contents and during this process they cause staining. They cause little strength reduction. Eg. Blue stain caused by *Penicillium* spp., *Ceratocystis* spp., *Cladosporium* spp and *Cladosporium* leads to deterioration of the wood quality.

2. **Decay:** The microbial decay takes place in the inferior core *ie.*, the heartwood of living trees. Decay process takes place in wood when the fibre saturation point of wood is around 28% moisture. Due to temperature and humidity fluctuations formation of a film of water in the cell lumen takes place and this leads to decay process. Sapwood decay is usually associated with large wounds. Decay takes place in two stages as primary decay where the wood attack is visualized as a slight to marked colour change from normal and advanced decay where the wood becomes more and more affected until it is finally changed in appearance and structure, leading to the destruction of the tissues. The decays continue to develop under favorable conditions and trees after conversion leads to rapid deterioration of the timber.

IV. Decay Based on the Position of Fungal Fruiting Bodies

Decays can often described according to the position of fungal fruiting bodies in the tree as root rot, butt rot or stump rot or on the type of wood affected as heart rot or sap rot.

i. **Heart rot:** Decay of living trees. Eg. *Phellinus pini* (*Fomes pini*), *Ganoderma applanatum, Fomes annosus*

ii. **Sap rot:** Decay of dead trees or wood products. Eg. *Coriolous versicolor*

iii. **Top rot:** Fungal fruiting bodies occur on top of tree. Diseases do not spread by roots. Eg. *Fomes* sp.

iv. **Root and butt rot:** Fungal fruiting bodies occur on base or on roots. Disease spread by roots. Eg: *Ganoderma lucidum, Armillarea mellea, Polypous tomentosus Heterobasidium annosum.*

v. **Slash rot:** Affects and decays branches, stubs and roots of dead trees. Many sap rotters behave as slash rotters (*Pycnoporus sanguineus*).

vi. **Products decay / rot:** A group of slash rotters attack the wood logs, wood product and cause deterioration. Eg. *Poria incrassata*

Factors Affecting Decay Process

The decay of wood is highly influenced by some factors that decides the progress and severity of the decay

1. **Age of the tree:** The age of the tree is another deciding factor for fungal invasion. Aged and matured trees are more prone to attack by heart rot with age.

2. **Temperature:** The atmospheric temperature influences the fungal decay process. The optimum range of temperature for the fungi ranges from 17- 30°C. The growth and enzymatic activities of the fungi are adversely affected by the very low and very high temperature.

3. **Air- moisture balance:** The balance between air humidity and wood moisture is another factor for the development of decay. Being a hygroscopic material, wood can absorb/ desorb and exchange water molecules with the surrounding air. Higher RH with 40 -70% wood moisture content favours decay. The wood moisture if maintained below 20 % will be free from infection by decay fungi.

4. **Wood pH:** Activity of decay fungi is optimum at pH 6 to 7. Progress of fungal invasion is affected at lower or higher pH

5. **Wounds / Injury:** The tree wounds and injuries caused by frost and fire are the main entry point for many of the fungi and insects which cause different types of rots on the forest plantations.

6. **Natural Resistance of the wood.** Fungal invasion is based on the natural characters of tree species such as thickness of the bark, nature of sap wood, nature of heart wood, pH of substrate and whether soft wood or hard wood. The natural exudates especially phenols and coumarines formed during development of the heart wood are deposited in the cell wall that confer resistance to fungal invasion.

7. **Physical barriers of wood:** Due to fungal invasion, development of tyloses in the infected tissue or formation of calluses above and below the infection points will lead

to compartmentalization of infected tissue from healthy tissues prevents further spread.

Prevention of Decay

In timber, the fungal pathogen attack after felling and conversion of wood may initiate from

i. The fungus present in the living tree may multiply in the dead wood.

ii. Fresh infection on the dead wood after felling and conversion.

1. Prevention of decay after felling

i. The felled and converted logs should be transported from forest immediately.

ii. The wood must be kept dry; do not expose to alternate wet and dry conditions.

iii. Moisture content of wood must be maintained below 20%

iv. Suitable protective chemicals may be applied at different stage of the conversion process: apply immediately after felling, after conversion and before storage. In warm and humid weather, the preservatives should be applied within 48 hrs after falling as stain causing fungi may infect quickly the surface of the log.

2. Prevention of decay during storage

i. The wood logs of species whose bark are not resistant to fungal attack may be removed thereby eliminating the risk of infection.

ii. Wood stored in water is free from decay as the oxygen can be reduced by submerging the wood in water.

iii. Piling of woods in racks preferably iron rather than directly on soil and allowing water sprays to keep wood moist prevents decay.

iv. Also temperatures below 10°C and above 93°C are not suitable for fungal growth, so maintaining required temperatures may prevent decay.

v. Main protective chemicals are: a mixture of pentachlorophenol in oil, a mixture of benzene hexachloride in oil, a sodium salt of pentachlorophenol mixed in water.

vi. After preservative treatment, the logs may be end- coated by applying thick, viscous bituminous preparation of asphalt or coal tar pitch. Alternatively, mud plaster can also be applied as end-coating. The end coating is useful in preventing infection of the fungi from any cracks left out in the chemically treated wood lots.

vii. In biological protection a relatively beneficial harmless microorganism is inoculated in to wood logs for affording protection against wood decay fungi. Fresh wood log can be sprayed with 4% solution of sodium fluoride or 5% ammonium bifluoride so as to induce the growth of Trichoderma viride which prevents the growth of decay fungi.

viii. Periodic checking of the logs to be done and any log showing the symptoms must be removed.

3. Prevention of decay of pulp wood

It is normally usual to store pulp woods for some time before is converted in to pulp. but it is advisable to remove pulp wood immediately from forests. Chemical pulp is used quickly within a few months. But mechanical pulp is stored for a year in open during which the stain fungi, moulds and affect the pulp and cause discolouration that affects quality as the pulp particles are bound together producing a lumpy speckled paper. Also the wood destroyers make the wood fibers brittle so that they are broken and much pulp is lost in water. Bamboo culms may be stored in huge piles in the yards of paper mills which is prone to attack by decay fungi. Decay is severe in humid and tropics where high rainfall is there. Numerous holes occur in the fibre that reduces the unbleached pulp wood quality. In bleached pulping, the yield gets reduced. Application of sodium pentachlorophenate at 0.9 to 1.8 kgs per ton of dried pulp a spray will reduce deterioration by decay fungi.

4. Prevention of decay in chips

Wood used for pulping is stored in the form of chips which is also prone to decay during storage. Hence, storage humidity and temperature has to be maintained at optimum levels.

5. Prevention of decay in use

Wood that is used for indoors must be kept with moisture content below 20% by proper seasoning. Wood should be kept dry by painting, moisture proofing, adequate ventilation, use of treated and durable wood, providing chemical treatment, applying coal tar to exposed ends.

Preservatives for Prevention of Timber Decay

Wood preservatives: The wood decay can be eliminated by proper treatment with some chemicals- preservatives. The treatment usually helps in increasing the service life of the wood.

Properties of wood preservatives

- » A good preservatives should be highly toxic to the fungi
- » Low volatility
- » High resistance to leaching and more ability to penetrate deep into the wood
- » Non corrosive to metals
- » Safe to handle

Wood preservatives are classified by the type of application and based on the term of protection. Some preservatives have leach resistance and broad spectrum efficacy to protect wood that is exposed directly to soil and water. Other preservatives are having intermediate term of preservation and some do not withstand precipitation.

Types of preservatives

1. **Preservative oils:** Creosote is the very commonly used oil for preservation of wood used in railways and marine industry. It is toxic to many wood decaying organisms. It is low cost, highly toxic to fungi, easily applied.

2. **Chemicals soluble in water**: Water borne preservatives can withstand leach and get fixed. They are less corrosion to metals. The chemicals like Chromated copper arsenate (CCA), Ammoniacal copper zinc arsenate (ACZA), Alkaline quarternary compounds (ACQ), Copper azole, are used for preservation of woods.

3. **Chemicals soluble in organic solvents**: The chemicals like pentachlorophenol, zinc and copper napthenates come under this group. They have high degree of toxicity, they do not leach out

4. **Fungicides as preservatives**: Propicaonazole- Tebuconazole- Imidacloprid (PTI) can be used as it controls both fungal and insect attack

5. **Borates as preservatives**: They are unfixed water borne preservatives that can leach from treated wood and not much used.

Methods of application of preservatives

1. **Surface application:** This type of application is used for short term preservation of wood as it is applied on the surface of the wood.

2. **Wood Impregnation:** This method is useful for long term preservation of the wood, in which the chemical is impregnated into the wood either by open tank treatment under atmospheric pressure or by employing pressure to fill the cells of wood with the chemical.

3. **Boucherie method**: In this method, the sap wood is replaced with the preservative chemical.

Selected References

Borkar P, Doshi A, Navathe S. 2015. Mushroom diversity of Konkan region of Maharashtra, India. Journal of Threatened Taxa 7(10): 7625–7640.

Bruce A. 1992. Biological control of wood decay. IRG/WP/1531.

DIN 68800, Part 4. 1992. Wood preservation; control measures against wood-destroying fungi and insects. Beuth, Berlin

Mirko S. 1975. A Colour Guide to familiar Mushrooms and Fungi. Octopus Books, London, 126 pp.

Natarajan K. 1995. Mushroom flora of south India (except Kerala), pp. 387–397. In: Chadha, K.L. and S.R. Sharma (eds.). Advances in Horticulture, Vol. 13 - Mushroom. Malhotra Publishing House, New Delhi.

Natarajan K, Raman N. 1980. South Indian Agaricales - IX. Sydowia 33: 225–235.

Natarajan K. 1979. South Indian Agaricales V: *Termitomyces heimii*. Mycologia 71(4): 853–855.

Schwarze F. 1994. Wood rotting fungi: *Fomes fomentarius* (L.: Fr.) Fr. Mycologist 8:32–34

Thiribhuvanamala G, Kalaiselvi G, Parthasarathy S, Anusha B. 2017. Induction of lignolytic enzyme activities in different agro residues by the white rot fungi, *Pleurotus sajar-caju*. International Journal of Chemical Studies. 5(2), 89-94.

Thiribhuvanamala G, Kalaiselvi G, Parthasarathy S, Madhavan S, Prakasam V. 2017. Extracellular secretion of lignocellulolytic enzymes by diverse white rot basidiomycetes fungi. Annals of Phytomedicine. 6(1): 20-29.

Thiribhuvanamala G. 2017. Major diseases affecting pulp wood trees and their management. In: K.T. Parthiban, R. Seenivasan (Eds) Plantation and Agroforestry Pulpwood Value Chain Approach, Scientific Publisher, India.

Tiwari CK, Sharma N, Verma RK. 2005. Taxonomic over view of *Ganoderma lucidum* with special reference to Indian species. Indian Journal of Tropical Biodiversity 13: 57–72.

Tiwari CK, Parihar J, Verma RK, Prakasham U. 2013. Atlas of wood decaying fungi of Central India. Tropical Forest Research Institute, Jabalpur, MP, 166p.

Guide to Useful Books and Sources of Information on Tree Diseases

Publications are arranged under three general headings:

i. General

ii. Global

iii. Regional or country emphasis

I. General

This selection of books will help the reader learn more about pests and the more technical and scientific aspects of tree health. Some specifically emphasize forestry while others concentrate on insect pests and diseases of crop plants. The emphasis on a particular pest group is of less importance, when it comes to understanding biology, ecology, infection processes of pathogens and feeding habits of insects.

1. CABI crop protection compendium: global module. 2001. Wallingford, UK, CABI Publishing. The most comprehensive and up-to-date source of information on all groups of plant pests. Copiously illustrated and contains photos that can be cut and pasted to create data sheets. Few forestry species are included; the equivalent forestry compendium has information on tree pests but is principally a silvicultural guide. The CABI crop protection compendium is an excellent and affordable guide to symptoms and features of pest attack on plants. (www.cabicompendium.org)

2. Field and laboratory guide to tree pathology. R.O. Blanchard and T.A. Tattar. 1997. New York, USA, Academic Press. A standard guide to laboratory and field methods and procedures. Mainly concerned with fungal diseases although other non-insect pests are included. (www.academicpress.com)

3. Foliar symptoms of the nutrient disorders in the tropical shrub legume Gliricidia sepium. F.W. Smith and P.J. Vanden Berg. 1992. Division of Tropical Crops and Pastures Technical Paper No. 30. Indooroopilly, Queensland, Australia, CSIRO Publishing. Well illustrated, in colour. One of a series of three books that describes the effect of nutrient imbalances in important woody legumes based on experimental work in Australia; the other two books are on nutrient disorders in *Leucaena* and *Calliandra* species. (www.publish.csiro.au)

4. Forest health: its assessment and status. J. Innes. 1995. Wallingford, UK, CAB International. The standard text addressing all factors, including tree pests, that affect the condition and overall health of trees. It is both a scientific account of tree health studies from Europe, North America and other countries, and a practical guide. Selected problems are illustrated with black and white photos. (www.cabi.-publishing.org). Insect pests in tropical forestry. M.R. Speight and F.R. Wylie. 2000. Wallingford, UK, CAB International Publishing.

5. Plant pathologist's pocketbook. 3rd edition. J.M. Waller, J.M. Lenné and S.J. Waller. 2001. Wallingford, UK, CABI Publishing. A practical guide to laboratory and field practice. Contains advice on collecting samples, survey techniques and photography. Emphasis is on crop plants, but some information on trees. (www.cabi-publishing.org)

6. Principles of forest pathology. F.H. Tainter and F.A. Baker. 1996. New York, USA, John Wiley and Sons. Describes scientific aspects of tree pathogens based on examples and experiences in North America. (www.wiley.com)

II. Global

1. Compendium of conifer diseases. E. Hansen and K.J. Lewis. 1997. St. Paul, Minnesota, USA, APS Press. Main examples are taken from North America and Europe but also information on major conifers from other continents and information on disorders. All problems are illustrated by colour photographs, but are often too small to be practical. Good source of up-to-date knowledge on major disease problems. (www.shopapspresss.org)

2. Decline and dieback of trees and forests. W.M. Ciesla and E. Donaubauer. 1994. FAO Forestry Paper 120. Rome. A good introduction to abiotic diseases with a global overview; most examples are taken from Europe and North America. The photographs are in black and white and describe selected problems. (www.fao.org)

3. Diseases and pathogens of eucalypts. P.J. Keane, G.A. Kile, F.D. Podger and B.N. Brown. 2000. Indooroopilly, Queensland, Australia, CSIRO Publishing. A comprehensive scientific account from Australia; also includes information from other countries and their diseases. A reference book rather than a practical guide. Contains many colour photographs but their use for diagnostic purposes is limited. (www.publish.csiro.au)

4. An illustrated guide to the state of health of trees recognition and interpretation of symptoms and damage. Eric Boa, 2003. Food and Agriculture Organization of the United Nations, Rome. This guide helps readers to recognize symptoms of ill health in trees and understand their general significance. (http://www.fao.org/3/y5041e/y5041e00.htm#TopOfPage)

III. Regional or country Emphasis

1. ASIA:

Manual of diseases of tropical acacias in Australia, South-East Asia and India. K. Old, S.S. Lee, J. Sharma and Q.Y. Zi.. 2000. Bogor, Indonesia, CIFOR. First major account of tree pest problems reported from India to Australia where acacias have been widely planted for nascent forest plantation industries. The clear and helpful photographs of symptoms and pests will help in diagnosing problems. (www.cifor.org)

2. UNITED KINGDOM:

Diagnosis of ill health in trees. R.G. Strouts, R.G. and T. Winter. 2000. London, UK, The Stationary Office. An excellent book, practical with well-chosen and organized colour photos. Information comes from many different starting points: host tree, pest name, symptom or type of damage. Much effort has been expended in providing a clear layout, as well as succinct and informative text. (www.the-stationary-office.co.uk)

3. EUROPE:

Tree diseases and disorders: causes, biology and control in forest and amenity trees. H. Butin. Edited by D. Lonsdale. Oxford, UK, Oxford University Press. General text with an emphasis on Europe. Also describes individual diseases and disorders (nonpest problems) on a selected range of tree species. (www.the-stationary-office.co.uk)

4. NORTH AMERICA:

Diseases of trees and shrubs. W.A. Sinclair, H.H. Lyon and W. Johnson. 1989. Ithaca, NY, USA, Cornell University Press. A large format book with the most comprehensive collection of good and well-reproduced photographs. Although the selection of tree species is geared towards all aspects of forestry in North America, this book has wider applications in recognizing and interpreting symptoms. It is difficult to examine lists of diseases on individual tree species because of how this information is condensed and displayed in the index. However, the accounts of different diseases are arranged mostly by pest, and are clear and informative. (www.cornellpress.cornell.edu)

5. NORTH AMERICA:

Insects that feed on trees and shrubs. 2nd edition. W.T. Johnson and H.H. Lyon. 1991. Ithaca, NY, USA, Cornell University Press. An impressive and widely useful book that predated the companion volume, Diseases of trees and shrubs (see above). A separate index for insects, mites and other animals makes it easier to locate information. (www.cornellpress.cornell.edu)

6. USA:

Insects and diseases of trees in the South. Protection Report R8-PR16. United States Department of Agriculture - Forest Service, Southern Region. Includes both insect pests and diseases of commercial forestry species. (www.fhpr8.srs.fs.fed.us/idotis/insects.html)

7. INDIA

Forest Pathology: Principles and Practice in Forestry. B.K. Bakshi. 1976. Forest Research Institutes, Dehradun, India. It is a comprehensive and standard textbook of forest pathology in India. Elaborates about a basic concepts, disease development and types, seed pathology, forest disease control, forest plant quarantine and symptoms and management of tree diseases. (https://www.cabdirect.org/cabdirect/abstract/19771333955)

Government Agencies

1. Australia

Ensis Forest Biosecurity and Protection (Ensis joins Australia and New Zealand agencies).

2. Canada

Canadian Forest Service

- Forest health page
- Atlantic Forestry Centre, Fredericton, New Brunswick
- Great Lakes Forestry Centre, Sault Ste. Marie, Ontario
- Laurentian Forestry Centre, Sainte-Foy, Quebec
- Northern Forestry Centre, Edmonton, Alberta
- Pacific Forestry Centre, Victoria, B.C., has a very nice forest pathology site that focuses in part on root and butt rots important in the Northwest, and includes a super online diagnostic and disease guide: Common Tree Diseases of British Columbia
- Forest health research programs in Canadian Forest Service

British Columbia, Ministry of Forests, Forest Practices Code Guidebooks
Ontario Ministry of Natural Resources.

3. **Finland**

   » Finland (Forest Health Services) including basic facts, glossary, and an online diagnosis engine, no longer in English (Finnish)

4. **New Zealand**

   » Ensis Forest Biosecurity and Protection (Ensis joins Australia and New Zealand agencies)

5. **South Africa**

   » South African Tree Pathology Co-operative Programme

6. **Sweden**

   » Swedish University of Agricultural Sciences Forest Mycology and Pathology

7. **Switzerland**

   » Federal Institute for Forest, Snow and Landscape Research, Birmensdorf

8. **United Kingdom**

   » UK Forestry Commission Plant Health Page

9. **United Nations, Food and Agriculture Organization, forestry page**

10. **United States**

    » National Forest Health Protection page. Shortcuts to regional forest health pages:

11. **Colorado**

    » Colorado State Forest Service tree pest publications

12. **Minnesota**

    » Minnesota Dept. of Natural Resources tree care

13. **Oregon**

    - Oregon's Department of Forestry forest health page

Scientific Societies

1. American Phytopathological Society (APS) and APS Forest Pathology Committee
2. Canadian Phytopathological Society
3. International Society of Arboriculture
4. International Union of Forest Research Organizations – IUFRO divisions and working groups in pathology:
    » Pathology
    » Root and Stem Rots
    » Foliage, Shoot and Stem diseases
    » Vascular Wilt Diseases
    » Phytoplasma and Virus Diseases
    » Rusts
    » Disease/Environment Interactions in Forest Decline
    » Diseases of Tropical Forest Trees
    » *Phytophthora* Diseases on Forest Trees
    » Pine wilt disease
    » Parasitic Flowering Plants in Forests
    » Forest Health in Southern Hemisphere Commercial Plantations
    » Entomology (Several groups cover both insects and pathogens)
    » Impacts of Air Pollution and Climate Change on Forest Ecosystems
5. ORPATH International forest pathology mailing list hosted by IUFRO
6. Mycological Society of America
7. Quebec Society for the Protection of Plants
8. Society of American Foresters

9. North American Regional Groups
10. Commonwealth Forestry Association
11. Indian Society of Tree Scientists
12. Indian Society of Agroforestry

Universities, Extension and Other

1. Tree Pests and Disease course taught by William Livingston at University of Maine
2. Forest pathology on Facebook, by Jorge Alberto Cordero Martínez, Postgraduate Student in Phytosanitation (Plant Pathology Orientation) at the Colegio de Postgraduados-Campus, Montecillo, Texcoco, Mexico State (Mexico)
3. Diseases of Forest and Shade Trees course taught by Robert Blanchette at University of Minnesota
4. Web page of Thomas Harrington, Iowa State University
5. Web page of Dale Berghdahl at University of Vermont
6. The Bugwood Network at University of Georgia (online documents and photos)
7. Fact sheets from Pennsylvania State University
8. The Texas Plant Disease Handbook has good coverage of trees in a nice list of diseases, including descriptions and management recommendations.
9. The forest pathology section of the Plant Pathology Internet Guidebook has lots more forest pathology links.
10. Ornamental Pathologist's Internet Guide at North Carolina State University
11. University of Pretoria, Republic of South Africa, Forestry and Agricultural Biotechnology Institute (FABI)
12. Swiss Federal Institute of Technology, Zurich, Forest Pathology and Dendrology Group
13. Finnish Forest Research Institute, Ministry of Agriculture and Forestry of the Government of Finland
14. Forest and Nature Conservation Policy Group, Wageningen University, Germany

15. Association of Consulting Foresters of Australia (ACFA), Australia
16. UK's Tree Health and Plant Biosecurity Initiative (THAPBI)
17. The Forest Survey of India, Government of India
18. Indian Council of Forest Research and Education, Dehradun, India.

Web Sites on Specific Disease Types

1. Pitch canker in California
2. Tree bark and canker site at University of West Virginia
3. Root disease management guidebook in British Columbia
4. American Chestnut Foundation
5. Sudden oak death in California and in Oregon
6. Exotic Pests: Online conference on risks of exotic forest pests and their impact on trade
7. Exotic Forest Pest Information for North America (EXFOR)
8. Heartrots in Plantation Hardwoods in Indonesia and Southeast Australia
9. Saskatchewan Dutch Elm Disease Association

Journals

1. African Journal of Wood Science and Forestry
2. Agricultural and Forest Entomology
3. Agroforestry of Wasteland
4. Agroforestry Today
5. Annales Forestales (Anali za Sumarstvo)
6. Annals of Forest Research
7. Annals of Forest Science
8. Arboricultural Journal: The International Journal of Urban Forestry

9. Arboriculture and Urban Forestry
10. Australian Forestry
11. Canadian Journal of Forestry
12. Down to Earth (Indian)
13. East African Agriculture and Forestry Journal
14. Ecology Environment and Conservation (Indian)
15. Eurasian Journal of Forest Science
16. European Journal of Forest Research
17. Folia Forestalia Polonica
18. Forest Ecology and Management
19. Forest Pathology
20. Forest Review
21. Forests
22. Indian Forester
23. Indian Journal of Environmental Protection
24. Indian Journal of Forestry
25. Indian Journal of Marine Sciences
26. International Journal of Agriculture and Forestry
27. International Journal of Forest Research
28. International Journal of Forestry and Wood Science
29. Irish Forestry
30. Journal of Agroforestry System
31. Journal of Arboriculture

32. Journal of Forest
33. Journal of Forest Research
34. Journal of Forest Science
35. Journal of Forest Science
36. Journal of Forestry
37. Journal of Forestry
38. Journal of Sustainable Forestry
39. Journal of Tree Sciences
40. New Zealand Journal of Forestry
41. New Zealand Journal of Forestry Science
42. Research Journal of Forestry
43. Scandinavian Journal of Forest Research
44. Scottish Forestry
45. Silvigenetica
46. South African Forestry Journal
47. Swiss Forestry Journal
48. Sylwan
49. Tree Crop Journal
50. Tropical Forest Science
51. Wood Science and Technology

Indian Institutes

1. Autonomous institutes under Ministry of Environment, Forest and Climate Change
    - Tropical Botanical Garden and Research Institute, Trivandrum

» G. B.Pant Institute of Himalayan Environment and Development, Almora

» Indian Council of Forest Research and Education, Dehradun

2. Indian Council of Forest Research and Education based institutes and centres

Institutes

» Forest Research Institute (FRI), Dehradun

» Institute of Forest Genetics and Tree Breeding (IFGTB), Coimbatore

» Institute of Wood Science and Technology (IWST), Bangalore

» Tropical Forest Research Institute (TFRI), Jabalpur

» Rain Forest Research Institute (RFRI), Jorhat

» Arid Forest Research Institute (AFRI), Jodhpur

» Himalayan Forest Research Institute (HFRI), Shimla

» Institute of Forest Productivity (IFP), Ranchi

Centres

» Forest Research Centre for Skill Development, Chhindwara

» Forest Research Center for Livehood Extension, Agartala

» Forest Research Centre for Eco-Rehabilitation, Prayagraj

For Product Safety Concerns and Information please contact our EU representative GPSR@taylorandfrancis.com Taylor & Francis Verlag GmbH, Kaufingerstraße 24, 80331 München, Germany

Batch number: 10726366

Printed by Printforce, the Netherlands